5th International Conference on Chemical and Process Engineering (ICCPE 2016)

MATEC Web of Conferences Volume 69 (2016)

Jeju Island, Republic of Korea
23 - 25 May 2016

Editors:

Kadhim F. Alsultani
Kei Eguchi
H.J. Heeres
Jabbar H. Al-Baidhani
Yaxin Su
Sulaiman Al-Zuhair
Sheng-Tao Yang
Yu-Shao Chen

ISBN: 978-1-5108-2747-9

Printed from e-media with permission by:

Curran Associates, Inc.
57 Morehouse Lane
Red Hook, NY 12571

Some format issues inherent in the e-media version may also appear in this print version.

Printed by Curran Associates, Inc. (2016)

For additional information, please contact EDP Sciences – Web of Conferences
at the address below.

EDP Sciences – Web of Conferences
17, Avenue du Hoggar
Parc d'Activité de Courtabœuf
BP 112
F-91944 Les Ulis Cedex A
France

Phone: +33 (0) 1 69 18 75 75
Fax: +33 (0) 1 69 28 84 91

contact-edps@webofconferences.org

Additional copies of this publication are available from:

Curran Associates, Inc.
57 Morehouse Lane
Red Hook, NY 12571 USA
Phone: 845-758-0400
Fax: 845-758-2633
Email: curran@proceedings.com
Web: www.proceedings.com

TABLE OF CONTENTS

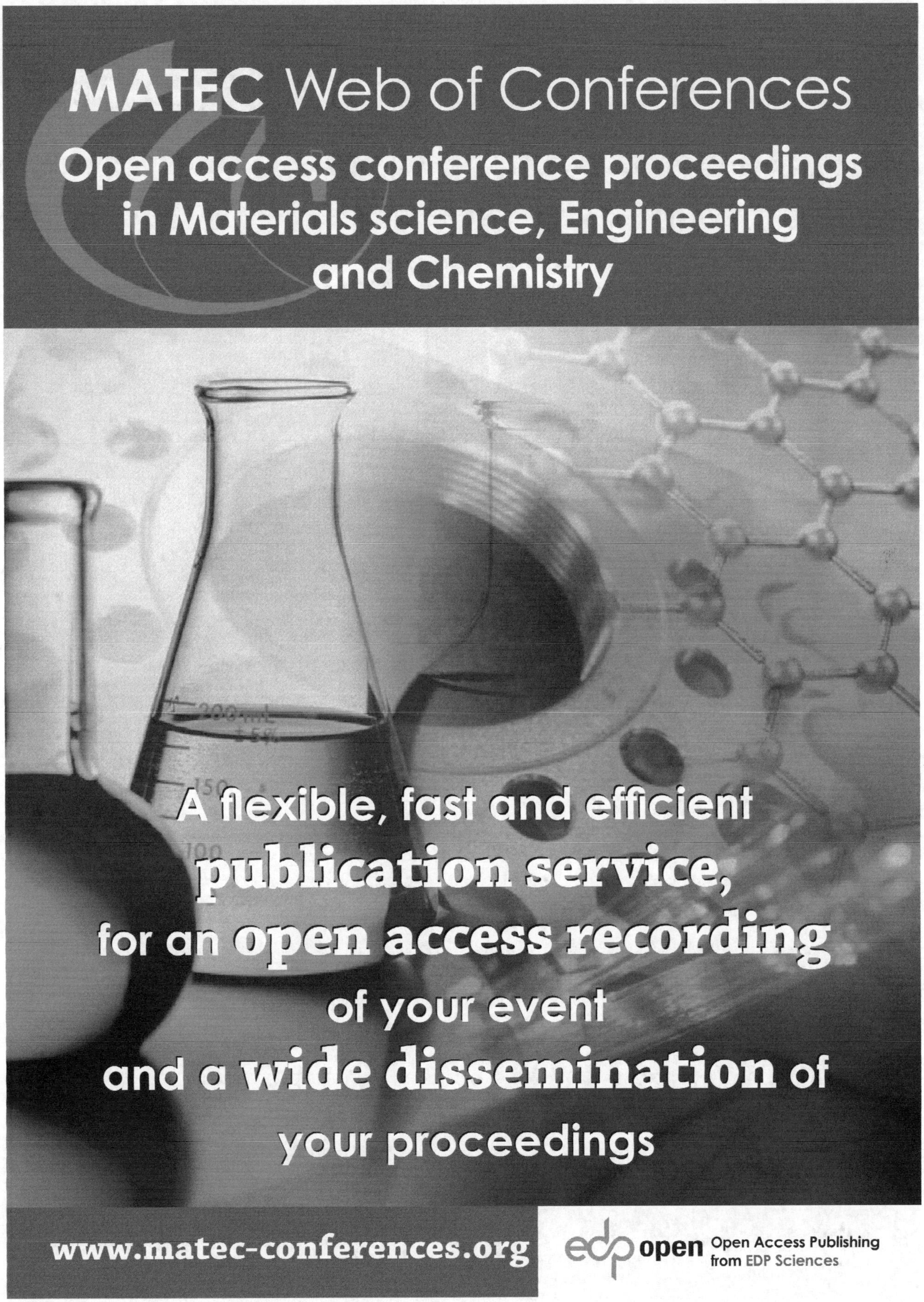
MATEC Web of Conferences
Open access conference proceedings in Materials science, Engineering and Chemistry
A flexible, fast and efficient publication service, for an open access recording of your event and a wide dissemination of your proceedings
www.matec-conferences.org
edp open Open Access Publishing from EDP Sciences

ISSN (electronic edition): 2261-236X

ICCPE 2016 Committee Members

Conference General Chair

Prof. Young Sun Mok, Department of Chemical and Biological Engineering, Jeju National University, Jeju, Korea

Technical Program Co-Chairs

Prof. Basim Abu-Jdayil, Chemical & Petroleum Engineering Department, UAE University, Abu-Dhabi, UAE

Prof. Kei Eguchi, Fukuoka Institute of Technology, Japan

International Technical Committees

Prof. Kei Eguchi, Fukuoka Institute of Technology, Japan

Prof. Basim Abu-Jdayil, UAE University, Abu-Dhabi, UAE

Prof. Arzu Akpinar-Bayizit, Uludag University, Bursa, Turkey

Prof. Jabbar Hammood Abdulnabi Al-Baidhani, Babylon University, Iraq

Prof. Kadhim F. Alsultani, College of Materials Engineering, Babylon University, Iraq

Prof. Sulaiman Al-Zuhair, UAE University, UAE

Prof. D V R Murthy, National Institute of Technology Karnataka, India

Prof. Yaxin Su, Donghua University, China

Prof. Sang Jin Park, Dongguk University, South Korea

Assoc. Prof. Sheng-Tao Yang, Southwest University for Nationalities, China

Assoc. Prof. Yu-Shao Chen, Chung Yuan University, Taiwan

Assoc. Prof. Mum Wai Yip, Tunku Abdul Rahman University College, Malaysia

Assoc. Prof. Robert Manurung, Institut Teknologi Bandung, Indonesia

Assoc. Prof. Marwa Elkady, Egypt-Japan University of Science and Technology (E-JUST), Egypt

Dr. Khor Ee Huey, Curtin University Sarawak, Malaysia

Dr. Sukanta Kumar Dash, Pandit Deendayal Petroleum University, Gandhinagar, India

Dr. Berceste Beyribey Price, Yildiz Technical University, Istanbul, Turkey

Dr. Siu Hua Chang, Universiti Teknologi MARA (UiTM) Malaysia

Dr. Mustafa Ahmed Rijab, Technical Institute of Baquba, Mechanical Department, Iraq

Dr. Nur Hidayati Othman, Universiti Teknologi Mara (UiTM), Malaysia

Preface

On behalf of the ICCPE organizing committees, we welcome you to the 2016 5th International Conference on Chemical and Process Engineering in Jeju Island, Republic of Korea during May 23-25, 2016. ICCPE 2016 is a conference dedicated to issues related to chemical engineering fundamentals, chemical engineering educational challenges and development, chemical reaction engineering, chemical engineering equipment design and process design, thermodynamics, interfacial & colloidal phenomena etc.

The major objectives of the conference is to bring together scientists, engineers, researchers from both the academic and industry to exchange and share their experiences and research results, and discuss the practical challenges encountered and the solutions adopted in every aspect of chemical and processing engineering. We hope you will have a delightful and rewarding experience at the conference.

The proceedings of ICCPE 2016 contain 27 selected papers from the conference which are presented orally at the conference or via poster sessions, and provide up-to-date, comprehensive and worldwide state-of-the art knowledge in this field. Each contributed paper was blind peer-reviewed by international reviewers who were drawn from the conference committee members as well as other external reviewers invited from all over the world. The proceedings are divided into 8 chapters based on the different topics of the papers, namely, biochemistry, bioenergy & chemistry, physical chemistry, material chemistry, membrane chemistry, analytical chemistry, analytical chemistry, manufacturing technology.

We would like to thank the members of the program committees and reviewers. They have worked very hard in reviewing papers and making valuable suggestions for the authors to improve their work. We also would like to express our gratitude to the external reviewers, for providing extra help in the review process, and the authors for contributing their research results to the conference.

Prof. Byoung Ryong Jeong
College of Agriculture & Life Science,Gyeongsang National University, Korea
ICCPE Organizing Committees
May 20, 2016.

Statement of Peer review

In submitting conference proceedings to MATEC Web of Conferences, I certify to the Publisher that I adhere to the **Policy on Publishing Integrity** of the journal in order to safeguard good scientific practice in publishing.

1. All articles have been subjected to peer review administered by the proceedings editors.
2. Reviews have been conducted by expert referees, who have been requested to provide unbiased and constructive comments aimed, whenever possible, at improving the work.
3. Proceedings editors have taken all reasonable steps to ensure the quality of the materials they publish and their decision to accept or reject a paper for publication has been based only on the merits of the work and the relevance to the journal.

Title, date and place of the conference

2016 5th International Conference on Chemical and Process Engineering (ICCPE 2016), May 23-25, 2016, Jeju Island, Republic of Korea

Proceedings editor(s):

Prof. Kadhim F. Alsultani, Babylon University, Iraq

Prof. Kei Eguchi, Fukuoka Institute of Technology, Japan

Prof. H. J. Heeres, University of Groningen, the Netherlands

Prof. Jabbar H. Al-Baidhani, Babylon University, IRAQ

Prof. Yaxin Su, Donghua University, China

Assoc. Prof. Sulaiman Al-Zuhair, UAE University, UAE

Assoc. Prof. Sheng-Tao Yang, Southwest University for Nationalities, China

Assoc. Prof. Yu-Shao Chen, Chung Yuan University, Taiwan

Date and editor's signature

2016-6-26

MATEC Web of Conferences **69**, 01001 (2016) DOI: 10.1051/matecconf/20166901001
ICCPE 2016

Effect of rich Selenium Mushroom Polysaccharides on Hyperlipidemia Metabolism and Lipoprotein Lipase Activity of Tests Mice

Yue Li-hong[1], Wang Yu-jun[1], Yu Jing-ni[1] and Zhang Yue-hua[2,a]

[1]*College of Life Science, Jiamusi University, China*
[2]*College of Science, Jiamusi University, China*

Abstract. The Selenium-Polysaccharides from Rich selenium Chemical explicitly defined the Edible Fungus (Se-EF) in Boiling water extraction by alcohol precipitation, analysis indicated that the there were main ingredient comprises xylose, dextrose, rhamnose, galactose, and arabinose. First investigate lipids Metabolic and lipoprotein lipase (LPL) activated in the selenium-rich -EF rich in cholesterol and prospects by feeding rich foods (PRD). The purpose of this study is to explore the prospect 1st fed cholesterol-enriched diet (PRD) of KM mice selenium -EF on cholesterol metabolism and lipoprotein lipase (LPL) activated of antioxidant capacity. In addition, studies of the relationship between atherosclerosis and total antioxidant capacity index (AI) and LPL activity (TAC). The results showed that selenium polysaccharide significantly reduce TC serum total cholesterol and LDL-C concentrations comparatively in with the CK control groups. Furthermore, orally administered of polysaccharide selenium notable enhance pay, lipoprotein lipase activity and reduce MDA levels, and artificial intelligence. These results indicate that the beneficial affects of selenium on -EF hypercholesterolemia precautions. In addition, oral administration of polysaccharide selenium notable enhance TAC and LPL activity in the meantime reduce MDA levels, the same as AI. These results indicate that Se-EF has a beneficial effect on the prevention of hyperlipidemia. These results indicate that the beneficial affects of selenium on -EF hypercholesterolemia precautions.

1 Introduction

The numbers of people suffering from hyperlipidemia, cardiovascular illness, rectally cancer was straight up in developing China. In medical, surgical, and endovascular treatment of tremendous progress, the significant number of intervention Implementation Constantly to grow each year. While Fatty and high protein composition is the main factors of lipid metabolism. It should not be ignored without food [1] skimmed part of. Statistics show that there is black fungus as dietary supplements May effectively reduce the incidence of cardiovascular and cerebrovascular. However, there are few papers on the direction of hyperlipidemia, atherosclerosis index and the lipoprotein lipase so far. In the past ten years, Lots of attention concentrated the metabolism of and atherosclerosis. In the past decade, A lot of attentions has concentrate on the affect of LPL on lipid cholesterol and atherogenesis [2]. Inhibition of LPL activity may Enhance TG and Reduction HDL and cholesterol (HDL-C) concentration. This theory is well known, it is assumed in the oxidation- reduction state. Changes due to LPL activity change. Therefore, in this study, it is assumed to that antioxidants may directly or indirectly affect LPL activity by varying oxidation and reduction state as the cells to studied the total antioxidant- reducing ability, cholesterol activity, MDA level, As well as other indicators, -EF effects And high body mass trumpet, the levels of Intake of high-fat foods, in vivo food conversion efficiency, lipid and KM mice atherosclerosis index [3]. Moreover, the Ratio of TC-(HDL-C) and (HDL-C) is the formula is Al, has been the risk of cardiovascular illness, like increased hazard in proportional to the increase in the proportion of directly related. For this perspective, the association between artificial intelligence and LPL activity and overall anti-oxidation ability (TAC) were evaluated between [4]. Therefore, the conclusion will promote know- ledge of black fungus cardiovascular benefit, and to explore its prevention in functional foods and health products for cardiovascular disease / therapeutic applications.

2 Materials and methods

2.1 Review stage

The method of Se-A. Auricular polysaccharide material was preparation. The basic components of the Se-A. Auricular material corresponding has been measurement. The total dietary fiber content by enzymatic method and gravimetric method has been measuring. The determination of total fat content was gas chromatography technique. Determination of nitrogen

[a] Corresponding author: zhangyaohua_2008@ 163·com

performed done according to the method established by using conventional methods using Bovine serum albumin (BSA) as Reference Sample. The nitrogen is converted to protein mass obtained, the correlation coefficient was calculated 6.25 [5]. The samples of A. auricular artificial was cultivation, in BOBABU Biotech Co. Ltd., Jiamusi City, Heilongjiang Province, China. A. auricular samples at 60 ℃, initial oven 2 hours, and on the ground with electric grinding mill and passed through a 40 meshes and 60 mesh sieves. A. auricular containing lyophilized carbo- hydrates (grams per 100 g dry weight) of 80 (comprises 54 grams overall dietary fiber), 8.0g Raw protein, fat 1.3g, ashes 8.7. Powder samples 200.0 g in 3.0 L alcohol extraction for 48 hours at room temperature, low molecular weight fat components and filtration removed through water-soluble substances Xin-Hua 2th paper to remove small molecular weight components of the lipid composition and insoluble's.

This process is performed three times. The filtered product was dissolved in distilled water and transferred to a 3 liters reflux apparatus purified at 100℃ condition for repeated two times. Extracts were filtered and concentrated under reduced pressure device, combined with precipitate was extracted with alcohol to obtain the polysaccharide was repeated 4 times. Obtained insoluble material was collected by centrifuge, washed by alcohol, freeze-dried to remove the volatile components, the crude polysaccharide obtained, namely 47.8 g, then powdery product with an electric grinder for subsequent research. The resulting product collect by centrifugation, and washed with alcohol, ion exchange technology and low-temperature drying, to obtain 48.7 g of crude sugar, and then with an electric grinder to grind into a powder.

2.2 Composition analysis

The Total carbohydrate contents in Calculated using the 3, 5-dinitrosalicylic acid as internal reference standard phenol-sulfuric acid method. The glucuronide is as a standard sample, using the colorimetric determination of total urinate. Sample ashes conditions are: Step 1: 600℃, 1 hour. Step 2: 800 ℃ constant weight. 4 hours. Determination of the nitrogen content was in accordance with the foregoing method.

2.3 Animal experiment

All of standardized male KM-Mice (Experimental Animal Center of the 2th Harbin hospital, China) 5 days before the experiment domestication were being randomly divided into experimental group. At a temperature of 23 ±2℃ and 55±5% humidity conditions, alone in a separate cage with 12 hours of light and shade in breeding mice (form 8:00 to 20:00), at 24±1℃ of and at the temperature of 60±4 % humidity, the timing weighed. In the feeding period, In the food (as shown in Table 1) and Water and food without restrictions, free to feed 5 wks, The experimental KM-mice were divided into three groups of 12, The first group named control group (NG), The second group as high cholesterol group (MG), and The third group is the self-treatment group (EG). NG of KM mice fed normal diet, MG and EG of KM mice were fed cholesterol-rich foods diet.

Table 1. Chemical ingredients of sample diets (In per100 g dry weight).

Compositions	Content	
	RD	PRD
Carbohydrate	58	49
Protein	21.4	21.6
Fiber	4.9	4.9
Fat	4.2	4.2
P	1.2	1.2
Ash	8	8
Sodium cholic acid	---	0.2
Cholesterol	---	1
Lard	---	8.7
Ca	2.7	1.9

Experimental animal feed ingredients shown in Table 1. During the eight weeks, EG DEF of KM mice were administered by oral gavage using 120 mg / kg / day of body weight once a day, MG, NG and orally administered an equal volume of distilled water as a control. Experimental animal body weight was measured in accordance with S-EF fraction oral gavage dose alternate days. Daily food intake measurement ibid. Blood was collected and analyzed every week [6]. Efficiency of food intake of laboratory mice is calculated based on the following formats: Daily weight gain / daily food intake × 100. Experimental procedures are performed according to the experimental animals manual Harbin Medical University. (Harbin, China).

2.4 Experimental animal serum samples collected

Orbital venous plexus blood samples KM mice without anesthesia, quickly taken from sterile heparin-containing capillary, operating respectively in 4, 6 and 8 weeks, respectively. The experimental animals were transferred to a sterile 10 ml blood plastic tubes, 37℃ in 15 min, centrifuged for 8 minutes at 3600 rpm condition, the serum was separated. The serum samples of experimental animals placed in a refrigerator at -20℃, for serum total cholesterol, LDL-C, HDL-C and triglyceride concentrations were measured serum lipids, these indicators by enzyme kit methods.

2.5 TAC, LPL and Lipid per-oxidation determinations

TAC, LPL and Lipid per-oxidation determinations were correlation detection kit purchased from Shanghai enzyme- linked Biotech Company. All operational requirements and processes are measured in accordance with the instructions

2.6. Statistical analysis of experimental data

Statistical experimental data to each group 12 KM mice a computing unit for ± standard deviation were calculated. The design of the experimental data was using a one-way analysis of variance approach to statistical analysis in routine experimentation [7]. TAC relationship between AI and LPL activity by the Pearson product moment correlation coefficient was clarify and analyzed by SPSS16.3.

3 Results

3.1 The chemical composition of the experimental samples of polysaccharide

Chemical test samples S-EF compositions are listed in Table 2, composed of glucose, rhamnose, uronic acids, xylose, and sulfate were the main composition besides smaller amounts of arabinose, galactose, and mannose etc. Furthermore, there is a lot of money in the amount of ash product, which is caused by byproducts of sulfate and polysaccharides ironic ion binding composition. These various proteins may form with sulfated polysaccharide conjugate, was stable structure.

Table 2.Chemical ingredient of polysaccharide from Se-A. Fungus

Ingredient	Content Component
Arabinose	0.2
Ash	20.4
Mannose	1.4
Rhamnose	13.4
Sulfate	15.7
Total sugars	41.5
Galactose	0.7
Glucose	12.6
Nitrogen	1.8
Se	2.63×10^4
U. acids	19.7
Xylose	15.3

Table 3.Changes of weight and food intake of KM-Mice in tested periods.

Test Project	EG	MG	NG
Initial body wt (g/mouse)	22 ± 2.2	22 ± 2.4	24 ± 2.0
Ultimate body wt (g/mouse)	41 ± 2.1	41 ± 2.0	42 ± 1.8
Weight gain (g/mouse/d)	0.3 ± 0.06	0.5± 0.07	0.3 ± 0.04
Food intake (g/mouse/d)	3.4 ± 0.6	2.9 ± 0.8	3.6 ± 0.7
Feeding efficiency (%)	11.1 ± 1.4	13.1 ± 0.9	10.8 ± 1.2
Note: (P < 0.05)			

3.2 The efficiency in experimental animals body weight, food intake and feed

Experimental data show that those animals in Table 3, the normal control group and two experimental groups no significant difference in weight indicators, the same as The levels of daily food intake of the EG and MG diet groups. The levels of feeding efficiency of the EG and the MG were 11.2 and 12.3 g, respectively, compared with normal control group is relatively higher.

Table 4. Influences of S-EF on serum lipids in NG, MG and EG.

Serum lipids		NG	MG	EG
TG	W2	1.77 ± 0.15	1.38± 0.15b	1.68 ± 0.12a
	W4	1.73 ± 0.12	1.39± 0.16b	1.66 ± 0.16a
	W6	1.76 ± 0.14	1.41± 0.19b	1.69 ± 0.13a
	W8	1.78 ± 0.12	1.38± 0.17b	1.67 ± 0.15a
TC	W2	5.07 ± 0.48	10.93±0.72b	7.09±0.11ab
	W4	5.11 ± 0.45	11.75± 0.83b	7.65± 0.13ab
	W6	5.14 ± 0.52	11.32 ± 0.56b	8.18 ± 0.16ab
	W8	5.09 ± 0.49	11.58 ± 0.76b	8.23 ± 0.21ab
HDL-C	W2	3.85 ± 0.59	4.11 ± 0.71	3.96 ± 0.98b
	W4	3.76 ± 0.61	4.24 ± 1.13b	4.08 ± 1.12b
	W6	3.78 ± 0.76	4.38 ± 1.18b	4.12 ± 1.04b
	W8	3.82 ± 0.83	4.35 ± 1.14b	4.19 ± 1.26b
LDL-C	W2	0.35 ± 0.04	0.49 ± 0.09b	0.39 ± 0.05a
	W4	0.38 ± 0.07	0.52 ± 0.12b	0.42 ± 0.07a
	W6	0.37 ± 0.06	0.55 ± 0.08b	0.48 ± 0.08ab
	W8	0.37 ± 0.12	0.54 ± 0.11b	0.48 ± 0.09ab
Note: (a: P < 0.05 versus MG; b: P < 0.05 versus NG).				

3.3 Effect of S-EF in lipoprotein and serum-lipid concentrations

After experimental samples intervention for 2 week, the serum concentrations of HDL-C LDL-C, and TC, the Raise obviously about on, the same as TG content in the same manner. As shown in Table 4. The serum-lipid profile Remained relatively constant in the tested lasted in MG and NG groups. In the medium term, S-EF interventions can significantly reduce LDL-C and TC levels; there is relatively slight elevation over time more than eight weeks during the rearing period. However, significant differences were not observed between MG and EG in the concentration of C and HDL. In addition,

S-EF can regulate the content of TG to normal level, which shows subtle difference with NG.

3.4 Influences of S-EF levels Intervention in LPL activity

The Effects of s-EF Intervention in LPL activity Illustrate the Function of S-EF in that LPL activity of PRD-fed test group as shown in Fig. 1. LPL activity of PRD-fed KM mice in the serum was significantly reduced. S-EF treatment significantly inhibited PRD induced negative changes, NG comparison with experimental data show that significant differences do not exist. However, these data suggest TAC data relating to indicators associated with AI, AI is directly related to the oxidative stress.

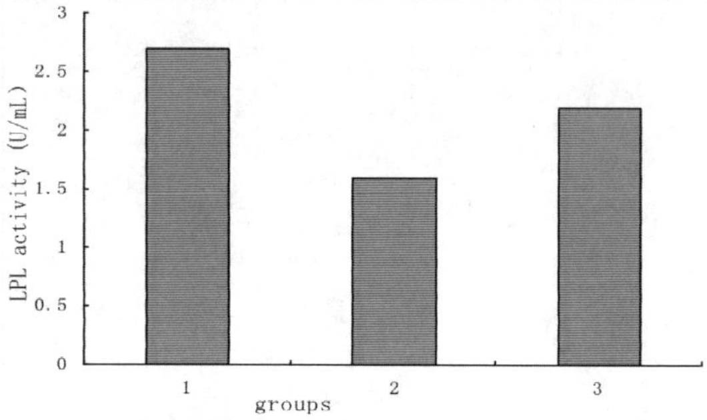

Figure 1. Compare of EF in LPL function of PRD-fed groups

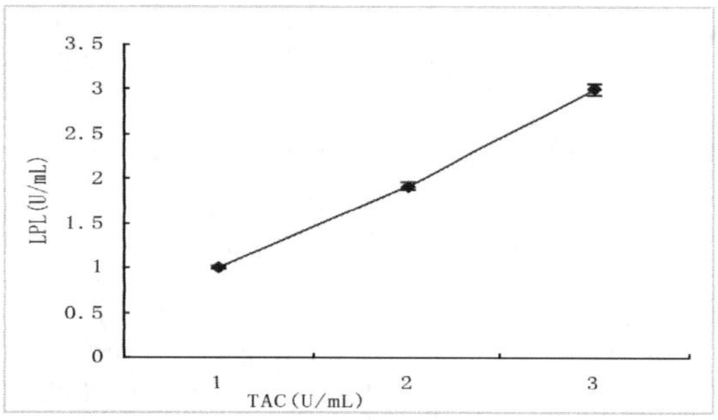

Figure 2. Correlation between TAC and atherosclerotic index.

3.5 Correlation between TAC and the AI

The relationship between AI and TAC was exposed by correlation analysis (Fig. 1). The analysis indicate a inverse correlation (Fig. 2). As a result, due to lower oxidation resistance, AI was relatively high in the series.

3.6 Activity of compare relationship between TAC and LPL

In Fig. 1, the correlation between the activity of LPL and TAC is represented. Positive correlation between LPL and TAC activity was discover. The correlation of TAC and LPL activity demonstrate that the change of redox state played anpositive function in the transform of LPL activity.

3.7 Effect of S-EF for MDA content

To investigate whether oxidative stress occurs in NG-fed KM mice, to satisfy of MDA was measured in these laboratory animals. The tests dates are disclosed in Table 5, to compare in MG, there was a remarkable reduces to compare MDA Gradation with the serum indicators of EG.

Table 5. Effect of EF on the AI of PRD-fed groups compared with control group. groups NG, MG and EG

Groups	NG	MG	EG
AI	0.35 ± 0.11	1.76 ± 0.21b	0.39 ± 0.07ab
TAC (U/mL)	2.54 ± 0.09	0.89 ± 0.04b	2.21 ± 0.05ab
LPL (U/mL)	2.79 ± 0.15	1.29 ± 0.09b	2.29± 0.12ab
MDA(nmol/mL)	9.99 ± 0.89	29.89 ± 2.38b	17.12 ± 1.21a

Note: a: $P < 0.05$ about MG; b: $P < 0.05$ about NG.

4 Discussion

S-EF Constraints therapy the downside diversification produced by PRD. it is consistently recognized that depression the level of high serum cholesterol administer to the prevention of atherosclerosis. National will take additional measures to decrease the concentration of serum cholesterol through adjusting High cholesterol food structure in their daily lives. Latest research achievements on human and animal subjects were indicated that a Diet increased fungus and other fungi foods have significantly lower blood lipids clinical effect [8]. These tests investigated the influence East Asian food (black fungus) polysaccharide on serum and blood lipid of mice is very clear intervention. In the research, the levels of HDL-C and LDL-C in serum of MG were differs notable with NG. This may be a stress response caused by high cholesterol levels in KM mice, which the correlation is significantly higher than of serum HDL-C and LDL-C. Which the Metabolic homeostasis is altered, the concentration of HDL-C and LDL-C in serum can be severely altered to form a hypercholesterolemia, even atherosclerotic sis.

This research suggests high cholesterol food intake and weight gain is insignificant in the diet were significantly different from that of the diet fed mice. The results showed that the contents of S-EF and TC ($P < 0.05$) in serum were significantly lower than that in serum LDL-C ($P < 0.05$).

It is also indicated that s-EF decreased serum TC may be due to decreased serum LDL-C. Results were consistent with previous findings. Based on the above research foundation, we think that S-EF is the main component of to reduce the serum TC concentration [9]. In addition, the balance of triglycerides also has the aforementioned studies show significantly increased risk of coronary artery disease and other heart diseases. The test data reveals that the activity of LPL TAC was positively correlated with, and I was a negative correlation TAC. Correspondingly it is whereas the AI was negatively correlated with TAC. This may be an important role in the reduction of AI. Lipoprotein has been a glycoprotein Change intervention fat into the peripheral tissue energy [10]. It frees fatty acid separation from TGs current in chylomicrons and unusual small density lipoprotein. Our research represents that LPL activity possibly was changed from the S-EF. In addition, according to our results, clear liquid level of MDA in magnesium was higher than that in NG. It indicates that the hypercholesterolemia may increase oil peroxide metabolism. It comes down to the hyper -cholesterolemia perhaps raise the small pieces, polymorphic nuclear cells, white blood cells and endothelial cells cholesterol content [11]. Thus MDA production may initiate a series of reactions and increase the oil peroxide route [12]. Our results indicated that S-EF oral prevented perhaps that hypercholesterolemia may cause the elevation of MDA. The conclusion results showed that, the whole weight of 120 mg/kg/d S-EF in along with the increment of diet management major stimulus in LPL activity and TAC, in which PRD feed KM Mice significantly reduced the supernatant of TC and LDL-C concentration and suppress the progress of atherosclerosis.

5 Conclusions

Comprehensive experimental data revealed that 120 mg / kg / day of body weight target food intervention S-EF remarkable lower serum total cholesterol and LDL-C concentrations and inhibition of atherosclerosis progression PRD KM mice fed with the increase significantly stimulate LPL activity and TAC. However, it's about the publication of the S-EF research is quite limited. Moreover, anti S-EF effect of anti-hypercholesterolemia accurate effective components, site and mechanisms have not yet to be determined. Therefore, more research is needed to clarify the S-EF products on anti-hypercholesterolemia and atheroprotection further.

Acknowledgment

This work was financially supported by science and technology research programs of Jiamusi University of key research projects, Number: 12Z1201513.

References

1. E. Guillamón, A. García-Lafuente, M. Lozano, Edible mushrooms. Fitoterapia 81, 7 (2010)
2. R. L. Yang, G. Le, A. Li, J. Zheng, Y Shi. Nutrition. 22, 11, (2006)
3. E. S. Ferreira, M. A. Silva, A. Demonte, V. A. Neves Soy. Journal of medicinal food. 14,1-2 (2011)
4. R. Di Benedetto, L. Attorri and F. Chiarotti. Journal of agricultural and food chemistry. 58,9 (2010)
5. G. F. Pauli, T. Gödecke, B. U. Jaki and D. C. Lankin. Journal of natural products. 75,4(2012)
6. C. M. Waggoner. California State University. Long Beach (2011)
7. R. B. Matthew, B. T. Jewell, M. F. Anne, H. Jason, M. H. David, L. T. Benzinger, J. C. Morris and M. A. Beau. Neurobiology of aging, 35,4 (2014)
8. G. Chen, Y. C. Luo, B. P. Ji and W. su. Journal of food science. 73,6 (2008)
9. R. L. Yang, G. Le, A. Li, J. Zheng, Y. Shi. Nutrition. 22,11 (2006)
10. S. Thangakrishnakumari, A. Nishanthini, S. Muthukumarasamy, V. R. Moha. Bennet (Asclepliadaceae) In Alloxan Induced Diabetic Rats. Asian Journal of Pharmaceutical and Clinical Research, 1,1 (2013)
11. Raghavan P R. Policosanol nanoparticles: U.S. Patent 8,722,093[P]. 2014-5-13.
12. M. Battino, P. Bullon, M. Wilsonand H. Newman. Critical Reviews in Oral Biology & Medicine. 10,4 (1999)

Comparison of two Agricultural Wastes for Phenol Removal Via Peroxidase-Catalyzed Enzymatic Approach

Tung Chiong [1,2], Ee H. Khor [1], Michael K. Danquah[1] and Sie Y. Lau[1]

[1]*Department of Chemical Engineering, Curtin University Sarawak, Malaysia*
[2]*Curtin Sarawak Research Institute, Curtin University Sarawak, Malaysia*

Abstract. Agricultural wastes of jicama and luffa skin peels were used as the source for peroxidase extraction. The extracted crude enzymes showed similar activities, 1.34U/mL and 1.22U/mL for jicama and luffa peroxidase respectively. These peroxidases were used to treat phenol under varying operating conditions of buffer pH, hydrogen peroxide concentration, enzyme volume and temperature. Jicama peroxidase demonstrated a phenol removal efficiency of approximately 90% at buffer pH 7, 1mM hydrogen peroxide using 1.5mL enzyme at 25°C. Luffa peroxidase required a higher dosage of hydrogen peroxide, and exhibited a removal efficiency of 84% at 8mM with other operating conditions same as jicama peroxidase. Jicama peroxidase is sensitive to pH change and more susceptible to thermal denaturation. Luffa peroxidase showed a better stability in terms of temperature.

1 Introduction

Phenol and its derivatives are present in various industrial effluents such as petroleum refineries, coal conversion, pulp and paper, resins and textiles. Phenolic compounds are mostly toxic and have been classified as hazardous pollutants [1], because they are harmful to microorganisms even at low concentrations. Therefore, these phenol-containing industrial effluents must be properly treated before they can be discharged into the receiving water bodies.

Conventional treatment methods for phenol-containing wastewater are generally microbial degradation [2], adsorption on activated carbon [3], and chemical oxidation such as ozonation [4] and Fenton reaction [5]. Although these methods are able to show high phenol removal efficiencies, some common drawbacks have been encountered. These include lengthy start-up procedure for microbial acclimatization [6], microbial growth inhibition due to high concentrations of phenolic compounds [7], intensive cost and energy for carbon regeneration [8], and disposal concern of reagents used during reaction process [8]. Due to the shortcomings of these conventional methods, enzymatic approach has emerged as an alternative for phenol removal from wastewaters, owing to its advantages which include low energy requirements, operation over a wide range of conditions, ease of process control, and minimal environmental impact [9].

The removal of phenols and other aromatic compounds from aqueous solution via peroxidase in the presence of hydrogen peroxide was first reported by Klibanov *et al.* [10]. Since then, horseradish peroxidase (HRP) became the most explored plant peroxidase, and its ability in removing various phenolic compounds from aqueous solutions are well documented [11]-[13]. Peroxidases from other sources such as soybean seed hulls, bitter gourd, cauliflower, potato and white radish have also been evaluated for their performances in treating phenol, and the reported results were encouraging [14]-[17].

It is notwithstanding that most of the plant peroxidases being studied are extracted from the edible parts of the plants, except soybean seed hulls. This will eventually cause competition with human food intake if the peroxidase extraction process is to be scaled-up to large capacity treatment. Hence, the present work aims to evaluate the performance of peroxidases extracted from agricultural wastes for phenol removal. Peroxidases were extracted from the skin peels of two locally available plants, such as jicama and luffa, which are non-edible and generally being discarded during food preparation. The performances of both crude enzyme extracts in phenol removal process were assessed and compared under varying reaction conditions.

2 Materials and methods

2.1 Extraction of peroxidase

Jicama and luffa purchased from local vegetable market were thoroughly washed with distilled water before their respective skin was peeled off. 100g of jicama and luffa skin peels was respectively blended with 200mL of 0.1M sodium carbonate buffer pH 10 for 30s, and then homogenized for 30min with constant stirring. The enzyme extract was then filtered through four layers of cheesecloth before being subjected to centrifugation at 4000 rpm at 4°C. The supernatant (crude extract) collected was sonicated and stored at 4°C until further use.

2.2 Batch treatment of aqueous phenol solution

The removal reaction was carried out by treating 1mM phenol solution with various dosages of enzyme volume in different buffer solutions. The enzymatic reaction was initiated by the addition of hydrogen peroxide (H_2O_2) into the reaction mixture, and the mixture was incubated for 24h with constant shaking to ensure maximum phenol removal. The efficiency of luffa and jicama peroxidases was studied in relation to the changes of pH of buffer solutions, H_2O_2 concentration, enzyme volume and temperature. The percentage of phenol removal was calculated based on equation (1).

$$\text{Phenol removal (\%)} = \frac{C_i - C_f}{C_i} \times 100\% \quad (1)$$

C_i = initial phenol concentration, C_f = final phenol concentration

2.3 Analytical procedures

Enzyme activity of peroxidase was measured using a colorimetric assay containing phenol, 4-aminoantipyrene (4-AAP) and H_2O_2. This assay is a modification of that developed by Wu *et al.* [18] in which the assay mixture consists of 250μL of 9.6mM 4-AAP, 100μL of 100mM phenol, 100μL of 2mM H_2O_2, 450-500μL of 100mM phosphate buffer (pH 6.0) and 50-100μL of enzyme solution. Prior to significant substrate depletion, activity was proportional to the rate of formation of a coloured product which absorbs light at a peak wavelength of 510nm with an extinction coefficient of 7100L/mol.cm based on the conversion of H_2O_2. One unit of activity is defined as the number of micromoles of H_2O_2 consumed per minute at pH 6.0 and 25°C.

Phenol concentrations were determined colorimetrically using 4-AAP and potassium ferricyanide in an alkaline buffer medium. Phenolic compounds react with 4-AAP under alkaline conditions to yield an intermediate species which is oxidized in the presence of the potassium ferricyanide reagent. The resulting compound is a quinone-type dye which absorbs light at 510nm. The colour intensity is linear with respect to phenol concentration, provided that this concentration does not exceed 0.1 mM in the cuvette [19]. The absorbance was measured at 510nm after 5 minutes.

3 Results and discussions

3.1 Enzyme activity of peroxidase

The extracted crude enzymes from the skin peels of jicama and luffa exhibited similar enzyme activities, being 1.34U/mL and 1.22U/mL respectively. The crude extracts were used as produced without any further purification.

3.2 Effect of pH

The effect of pH on phenol removal was conducted by using various buffer solutions ranging from pH 4 to 9 at 25°C. The buffer solutions used were acetate buffer (pH 4 to 5), phosphate buffer (pH 6 to 7) and borate buffer (pH 8 to 9). The dependence of phenol removal efficiency on pH is as shown in Fig. 1.

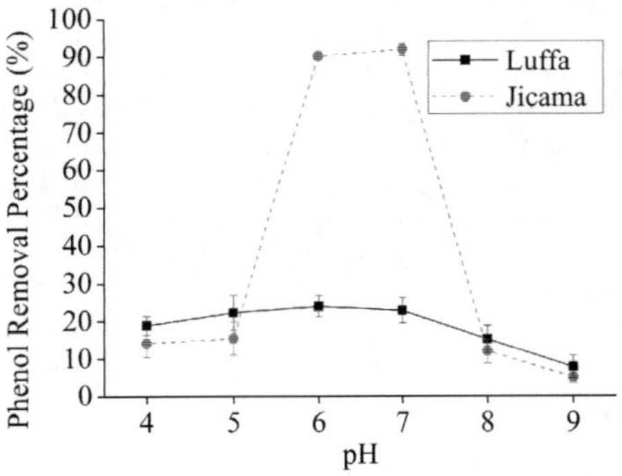

Figure 1. Effect of buffer pH on phenol removal (experiment conditions: 1.5mL enzyme, 1mM H_2O_2 at 25°C).

When 1mM phenol solution was treated with jicama peroxidase, removal efficiency of approximately 90% was observed at pH values 6 to 7, with the optimum occurring at pH 7. Removal efficiency decreased to less than 20% below pH 5 and above pH 8. This is a result of variations in enzyme protein structure in response to varying pH. The enzyme molecules undergo structural modifications as a result of protonation and hydroxylation effects which could potentially obscure the enzyme active sites before causing denaturation and permanent loss of functionality. Moreover, the decrease in removal efficiency at pH above 8 could be attributed to the formation of phenol conjugated base since the pK_a of phenol at 25°C is 10. This conjugated basic form does not permit the phenolic compounds to act as hydrogen donors, thus hindering binding onto the surface of the enzyme active sites.

Luffa peroxidase, on the other hand, showed low phenol removal efficiency of less than 24% over the pH ranges being evaluated. Though enzyme activities of both luffa and jicama were similar as mentioned earlier, this finding suggested that luffa peroxidase might require higher concentration of H_2O_2 for oxidation of its enzyme into catalytically active form which is capable of reacting with phenolic compounds [20].

Previous studies on phenol removal catalyzed by horseradish peroxidase (HRP) demonstrated optimal operating pH at pH 8 [12], which was slightly basic. Another work by Wright and Nicell [14] showed that nearly complete removal of phenol was observed over a pH range of 5 to 9, with the maximum removal at pH 6 by using high dose of soybean peroxidase (SBP). The optimum operating pH for jicama peroxidase was narrower than SBP, implying that it is more sensitive to pH change. All subsequent enzymatic reactions were conducted using phosphate buffer pH 7.

3.3 Effect of H_2O_2 concentration

Enzymatic reaction cannot take place in the absence of H_2O_2. H_2O_2 is required in peroxidase-catalyzed reactions to oxidize the native enzyme molecules (E) into Compound I (E_i) which then accepts an aromatic compound (AH_2) into its active site and carries out its oxidation to produce a free radical ($^{\bullet}AH$). The free radical is released from the catalytic site leaving the enzyme as Compound II (E_{ii}). Compound II oxidizes a second aromatic molecule, resulting in the release of a second free radical into the solution and returning the enzyme to its native state. The one-electron oxidation of aromatic substrate catalyzed by peroxidase is usually depicted by the following mechanism [21]:

$$E + H_2O_2 \rightarrow E_i + H_2O \quad (2)$$

$$E_i + AH_2 \rightarrow E_{ii} + {}^{\bullet}AH \quad (3)$$

$$E_{ii} + AH_2 \rightarrow E + {}^{\bullet}AH + H_2O \quad (4)$$

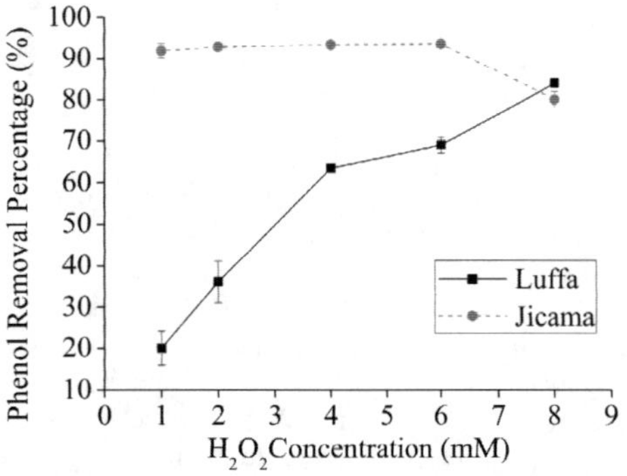

Figure 2. Effect of H_2O_2 concentration on phenol removal (experiment conditions: 1.5mL enzyme, buffer pH 7 at 25°C).

The effect of various concentrations of H_2O_2 on phenol removal by jicama and luffa peroxidases is as illustrated in Fig. 2. Phenol removal efficiency of more than 90% was recorded by jicama peroxidase over a range of H_2O_2 concentrations from 1mM to 6mM. When H_2O_2 concentration was further increased to 8mM, there was a decrease of about 10% in removal percentage. This could be attributed to enzymatic inhibition effect caused by H_2O_2. High concentrations of H_2O_2 inhibit peroxidase catalytic activity by irreversibly oxidizing the enzyme ferriheme group which is vital for peroxidase catalysis [22].

A different trend was exhibited by luffa peroxidase in regards to H_2O_2 concentration. When H_2O_2 concentration was increased, phenol removal percentage also increased. For the range of H_2O_2 concentrations being studied, the highest phenol removal efficiency of 84% was observed at 8mM. The reason that luffa peroxidase requires a much higher dosage of H_2O_2 for better performance in phenol removal could be due to the difference in the structure of its peroxidase molecules' active sites. The result obtained also suggested that luffa peroxidase is less susceptible to H_2O_2 inhibition, a phenomenon commonly suffered by peroxidase-catalyzed enzymatic method. Subsequent experiments for luffa and jicama peroxidases were conducted using 6mM and 1mM H_2O_2 respectively.

3.4 Effect of enzyme volume

The effect of enzyme concentration on phenol removal is important as it has a significant bearing on the process economics. The effect of crude peroxidase dosage in phenol removal from aqueous solution was investigated for both jicama and luffa peroxidases under varying volumes of 1.5mL to 4.5mL, as shown in Fig. 3.

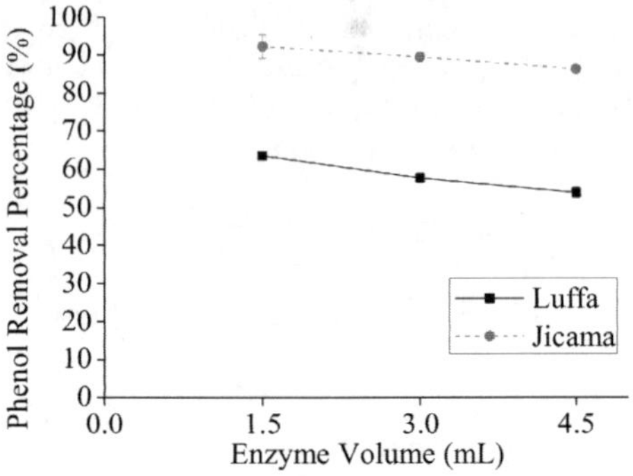

Figure 3. Effect of enzyme volume on phenol removal (experiment conditions: buffer pH 7, 6mM and 1mM H_2O_2 for luffa and jicama respectively at 25°C)

Both peroxidases exhibited similar trend, with removal efficiency decreasing when the enzyme volumes were increased. The efficiency of jicama peroxidase declined gradually from 92% at 1.5mL enzyme volume to 86% at 4.5mL enzyme volume. Luffa peroxidase showed a slightly higher drop, from 63% to 53% at 1.5mL and 4.5mL enzyme volume, respectively. Increase in enzyme volume means that there are greater amount of peroxidase molecules in the solution, and consequently increasing available active sites of enzyme for phenol binding. However, this scenario did not enhance the phenol removal efficiency as predicted. This could be a result of depletion and/or insufficient H_2O_2 under increasing concentration of enzyme. The heme prosthetic group of peroxidase reacts with H_2O_2 in the first step of the catalytic cycle, thus, under depleting levels of H_2O_2, this affects enzyme activation. A higher dosage of H_2O_2 might be required to activate both peroxidases at higher enzyme volumes.

3.5 Effect of temperature

The effect of temperature on phenol removal was examined by incubating the reaction mixtures at various temperatures, ranging from 25°C to 70°C. As depicted in Fig. 4, jicama peroxidase attained a maximum phenol removal percentage of 92% at 25°C whereas luffa peroxidase showed a constant percentage averaging at 64%. Elevated temperatures were found to demonstrate a significant adverse effect on jicama peroxidase as compared to luffa peroxidase. When the temperatures were increased from 25°C to 40°C, jicama peroxidase suffered a drastic drop in removal efficiency from 92% to 41%, which was more than half of the value. Its removal efficiency remained constant from 40°C to 60°C. A slight decrease in removal efficiency was again observed when the temperature was further increased to 70°C. Luffa peroxidase, however, is more stable towards thermal heat. Its activity can be preserved over a wider temperature range. This is shown by its nearly constant removal efficiencies over the range of 25°C to 70°C. Higher temperatures neither enhance nor inhibit its activity in oxidizing phenol compounds.

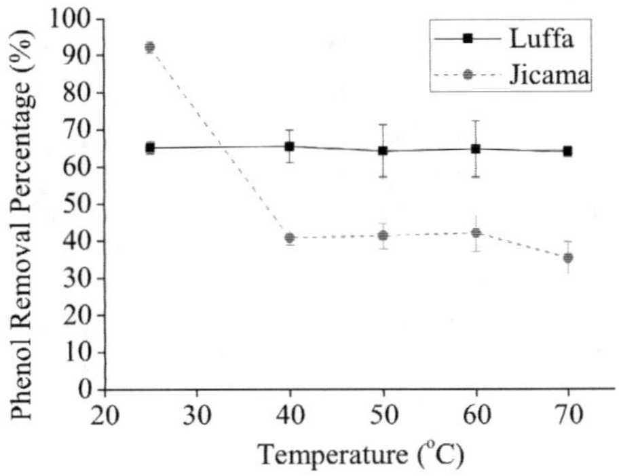

Figure 4. Effect of temperature on phenol removal (experiment conditions: 1.5mL enzyme, buffer pH 7, 6mM and 1mM H_2O_2 for luffa and jicama respectively)

The decline in efficiency by jicama peroxidase could be attributed to thermal denaturation. Under elevated temperature conditions, thermal denaturation of enzyme molecules causes loss of its active sites in catalyzing substrate molecules. The thermal stability of peroxidases is governed by the haem prosthetic group, which is released under elevated temperatures to form apoenzyme. The transient enzyme formed is less stable and more susceptible to thermal inactivation as compared to the native enzyme [23]. Higher temperature conditions distort the structure of the enzyme, causing a limited binding capacity of its active sites onto substrate molecules. The optimum operating temperature for both jicama and luffa peroxidases was selected at 25°C, in order to preserve peroxidase activities and save operating cost.

4 Conclusions

In the present study, both jicama and luffa peroxidases extracted from their respective fruit skin peels demonstrated promising performances in removing phenol from aqueous solution. Jicama peroxidase, though showing removal efficiency of about 90%, has a narrow operating pH range and is more susceptible to thermal denaturation. Luffa peroxidase requires a much higher concentration of H_2O_2 to activate its active sites before it can carry out the oxidation of phenol. Its enzyme activity is also well preserved under the elevated temperatures being studied. The findings from this work revealed the potential of jicama and luffa peroxidases in the application of phenol removal. Nevertheless, a more comprehensive study such as optimization of the operating conditions might be necessary to better understand the factors affecting the performances of these peroxidases in phenol removal.

References

1. N. Caza, J. K. Bewtra, N. Biswas, K. E. Taylor, Water Res. **33**, 3012 (1999)
2. A. Krastanov, Z. Alexieva, H. Yemendzhiev, Eng. Life Sci. **13**, 76 (2013)
3. S. Mukherjee, S. Kumar, A. K. Misra, M. Fan, Chem. Eng. J. **129**, 133 (2007)
4. K. Turhan, S. Uzman, Desalination **229**, 257 (2008)
5. Y. Yavuz, A. Savas Koparal, Ü. Bakir Ögütveren, Chem. Eng. Technol. **30**, 583 (2007)
6. T. T. Firozjaee, G. D. Najafpour, A. Asgari, M. Khavarpour, Chem. Ind. Chem. Eng. Q. **19**, 173 (2012)
7. W. Gernjak, T. Krutzler, A. Glaser, S. Malato, J. Caceres, R. Bauer, *et al.*, Chemosphere **50**, 71 (2003)
8. G. Busca, S. Berardinelli, C. Resini, L. Arrighi, J. Hazard. Mater. **160**, 265 (2008)
9. M. Gómez, M. D. Murcia, R. Dams, N. Christofi, E. Gómez, J. L. Gómez, Environ. Technol. **33**, 1055 (2011)
10. A. M. Klibanov, B. N. Alberti, E. D. Morris, L. M. Felshin, J. Appl. Biochem. **2**, 414 (1980)
11. M. Wagner, J. A. Nicell, Water Res. **35**, 485 (2001)
12. J. A. Nicell, J. K. Bewtra, N. Biswas, C. C. St. Pierre, K. E. Taylor, Can. J. Civ. Eng. **20**, 725 (1993)
13. S. Nakamoto, N. Machida, Water Res. **26**, 49 (1992)
14. H. Wright, J. A. Nicell, Bioresour. Technol. **70**, 69 (1999)
15. H. Ashraf, Q. Husain, Desalination **262**, 267 (2010)
16. A. N. Deva, C. Arun, G. Arthanareeswaran, P. Sivashanmugam, J. Environ. Chem. Eng. **2**, 1148 (2014)
17. J. Dec, J.-M. Bollag, Biotechnol. Bioeng. **44**, 1132 (1994)
18. J. Wu, K. E. Taylor, J. K. Bewtra, N. Biswas, Water Res. **27**, 1701 (1993)
19. M. Ghioureliotis, J. A. Nicell, Enzyme Microb. Technol. **25**, 185 (1999)
20. M. Hamid, K. u. Rehman, Food Chem. **115**, 1177 (2009)
21. J. A. Nicell, J. K. Bewtra, N. Biswas, E. Taylor, Water Res. **27**, 1629 (1993)
22. M. A. Duarte-Vázquez, M. A. Ortega-Tovar, B. E. García-Almendarez, C. Regalado, J. Chem. Technol. Biotechnol. **78**, 42 (2003)
23. J. P. McEldoon, J. S. Dordick, Biotechnol. Progr. **12**, 555 (1996)

Research of H5N6 Treatment by Comparing with H6N1 and H10N8 by Using Decision Tree and Apriori Algorithm

Sunghyun Kim[1], Hojoon Lee[1] and Taeseon Yoon[2]

[1]*Hankuk Academy of Foreign Studies, Natural Science Department, Yong-in, Republic of Korea*
[2]*Korea University, Computer Science and Engineering, Seoul, Republic of Korea*

Abstract. Since 2003, 608 people in 15 countries have infected with human-infectious AI viruses and 359 of them died. Especially, in China, H6N1 and H10N8 viruses were wide-spread and a lot of people were infected and died. Recently, H5N6 virus emerged in China and the number of patients has been increasing gradually. Therefore, this research compared amino acid strain of Matrix Protein, Hemagglutinin, Neuraminidase and Nucleoprotein of H5N6, H6N1 and H10N8, by using Decision tree and Apriori Algorithm, to figure out their similarity and devise the treatment. In result, Matrix protein and Nucleoprotein sequences of H5N6 were similar with those of H6N1 and H10N8. Therefore, this research concluded that the treatment targeting those proteins of H6N1 and H10N8 will be also effective to H5N6.

1 Introduction

Since 1878, when the first AI(Avian Influenza) was reported in northern Italy, a great number of chickens and ducks have infected with various AI viruses and been buried. [1] At the beginning of the outbreak, it was considered that only avian features could come down with the flu like its name, but now even humans are the host of AI viruses.

Since 2003, 608 people in 15 countries have infected with human-infectious AI viruses and 359 of them died. Especially, in China, the number of people who died due to viruses including H10N8 and H6N1, is overwhelming compared to other East Northern Asian nations. [2] As a considerable number of AI viruses in East Asia were from China, there is a strong possibility that human-infectious AI viruses arisen in China will spread to near countries. [3] In this situation, examining of H5N6, a new human-infectious AI virus in China, can contribute to the treatment of not only the virus but future-emerging AI viruses.

Thus, we are going to compare H5N6 with H10N8 and H6N1 by using Decision tree, Apriori Algorithm and discuss the treatment of it according to the result. [4]

2 Materials and methods

2.1 Avian viruses

Since 1990s, many influenza A viruses have circulated in poultry populations in China, which offer abundant gene pools for mutations. All of three viruses we use in this experiment are subtypes of influenza A viruses from China. [5]

2.1.1 H5N6

The first case of human H5N6 infection is a 49-year-old man from southwest China's Sichuan Privince in April, 2014. Since the initial case, nine human infections were reported, including the recent cases in 4 January 2016 and 11 January 2016 in China. [6]

2.1.2 H6N1

H6N1 viruses had circulated persistently in poultry, and continued to evolve and accumulate changes, increasing the risk of cross-species infection. Although H6N1 virus is one of the most common viruses in wild avian species, it has only infected one person, a woman in China, who even recovered. [7]

2.1.3 H10N8

Human infection with H10N8 was reported in China in December 2013 and January 2014. Those two patients showed signs of pneumonia, and died soon despite the administration of antibiotics and antiviral agents. Thanks to the later experiments, there is no more fatal infection case. [8]

2.2 Algorithms

2.2.1 Decision tree

The A decision tree is a way of algorithm specialized in classifying data and sequential decision problems. A decision tree uses explicit tree-shape graph so that many can easily read and understand.

A decision tree splits the remaining data and make common attribute to node. Then set groups into subsets that usually branches down with particular attribute. When expressed, the data is described mathematically for its generalization and categorization. It follows the form of (x,Y) = (x1, x2, x3,....,xk,Y). Vector x is composed of input variables and a dependent variable Y is the purpose should be understood, classified, or generalized. [9]

2.2.2 Apriori algorithm

Apriori is one of the usual algorithms of association rule. Apriori is a method for clarify correlation among data on the basis of frequency and general rules of data. Apriori has two steps to start processes. First, it finds groups of high frequency following the minimum support previously set. Second, figure out the association rules which satisfy value of confidence among the groups of high frequency. Then, all the groups of high frequency's subsets have high frequency also, and certain group can not have higher support than before regardless of a new added item. Therefore, Apriori works based on following sentence, for a start, it finds groups of high frequency among the groups whose size is 1. Then among the groups whose size is 2, 3, ... k, until it can't make the group, and get the answer. called bottom up approach. [10]

2.3 Experiment method

First of all, we collected nucleotide sequences of H5N6, H10N8, H6N1 viruses from NCBI, "National Center for Biotechnology Information". We designed experiment using Decision tree and Apriori algorithm in window5, window7, and window9 for each virus. By using those algorithms, we track out certain frequencies in Matrix Protein(M), Hemagglutinin(HA), Neuraminidase(NA), and Nucleoprotein(NP).

3 Result

3.1. Decision tree

Table 1. Result of hemagglutinin(window5)

	Pos1	Pos2	Pos3	Pos4	Pos5
H5N6	H		K E V	A	
H6N1			V		
H10N8	A N		E		

Table 2. Result of neuraminidase(window5)

	Pos1	Pos2	Pos3	Pos4	Pos5
H5N6	I		N S L T	L N I	V
H6N1	S	P K	I N Y V W	I G N Q	T Y K
H10N8	V		H S Q N V	Q P D G W N	T A G

Table 3. Result of hemagglutinin(window7)

	Pos1	Pos2	Pos3	Pos4	Pos5	Pos6	Pos7
H5N6				V	S		
H6N1			L	S D R T			S N
H10N8	S	T N		F	N S K E H	R T	

Table 4. Result of neuraminidase(window7)

	Pos1	Pos2	Pos3	Pos4	Pos5	Pos6	Pos7
H5N6				K			A G
H6N1			I	S V	S		D A E
H10N8	I		T	L G			E D

Table 5. Result of hemagglutinin(window9)

	P1	P2	P3	P4	P5	P6	P7	P8	P9
H5N6				N		I		V	
H6N1	R			T	G	E		V	
H10N8	A W I	G E	G	A E M	E H		L	D TI	

Table 6. Result of neuraminidase(window9)

	P1	P2	P3	P4	P5	P6	P7	P8	P9
H5N6	K				M	E SI			IS H
H6N1	N					D G		S	IF G H
H10N8	R E C						I	Q	S V Q G

By using decision tree, we track out rules of patterns with certain frequencies in M, HA, NA, and NP. A, C, D, E, F,

G, H, I, K, L, M, N, P, Q, R, S, T, V are names of amino acids. We use three viruses as experiment variables in every fold, and the results show both similarities and differences. We used the sequences for a 10-fold cross validation experiment, that is to say, we test 10 experiments Table1-6. When we gather extraction results, we first selected the rules of patterns of three viruses whose frequency rates higher than 0.750 and extracted at least two times. This is because degree under 0.750 is considered as almost unfixed rule patterns that are too unclear to be sure.

The decision tree reveals whole kinds of amino acids of HA and NA, so we can find lots of similar and different amino-acids. Also, there were proportional relationship between the lengths of the rule(window number) and the amount of particular rules of high frequency. Usually, the rules have frequency of 0.75, but window9 pos8=M and pos1=W have frequency of 0.833, for instance.

Furthermore, not as we expect, there was no rule in the experiment results in NP and M. Considering the characteristic of decision tree, we can see this result as different amino acids among these viruses did not have differences that are enough to be a key factor. Therefore we can see these viruses as similar species.

3.2 Apriori algorithm

A, C, D, E, F, G, H, I, K, L, M, N, P, Q, R, S, T, V are names of amino acids and the number means frequency. Star marks (*) beside amino acids in H6N1 and H10N8 mean that the rule is same with that of H5N6.

Table 7. Result of matrix protein(window5)

	H5N6	H6N1	H10N8
Rules	amino4=L 8 amino2=K 7 amino1=E 6 amino2=L 6 amino2=N 6 amino3=A 6 amino5=R 6 amino1=A 5 amino1=Q 5 amino1=S 5 amino2=A 5 amino3=G 5 amino3=M 5 amino3=Q 5 amino4=I 5 amino5=A 5	amino4=L 8 * amino1=A 6 amino1=E 6 * amino2=K 6 amino2=L 6 * amino3=A 6 * amino1=Q 5 amino2=S 5 amino3=G 5 * amino3=M 5 * amino3=Q 5 * amino4=I 5 * amino5=A 5 amino5=R 5 amino5=T 5 * amino5=V 5	amino4=L 8 * amino1=A 6 amino1=E 6 * amino2=A 6 amino2=L 6 * amino3=A 6 * amino5=A 6 amino1=Q 5 * amino2=K 5 amino2=T 5 amino3=G 5 * amino3=L 5 amino3=Q 5 * amino4=I 5 *

In M experiment Table 7, 9 of 17 rules of H5N6 were same with H6N1 and 8 were same with H10N8. Key factors are L(Leucine) and E(Glutamic acid). We discovered similar rate and feature in the experiment by window7 and window9.

In HA experiment Table 8, only 2 of 9 rules of H5N6 were same with H6N1, and none were same with H10N8. Even in the experiment by window7 and window9, we couldn't find any certain key factor..

Table 8. Result of hemagglutinin(window5)

	H5N6	H6N1	H10N8
Rules	amino3=L 16 amino5=N 14 amino1=L 13 amino5=A 13 amino3=K 12 amino2=E 11 amino3=S 11 amino4=N 11 amino5=I 11	amino2=N 14 amino1=S 13 amino3=K 13 amino3=L 13 amino4=G 12 amino5=E 12 amino1=L 11 amino4=N 11 * amino5=I 11 *	amino3=I 13 amino1=G 12 amino2=L 12 amino3=G 12 amino4=G 12 amino4=N 12 amino3=T 11

Table 9. Result of neuraminidase(window5)

	H5N6	H6N1	H10N8
Rules	amino1=I 13 amino3=T 11 amino1=N 10 amino1=S 10 amino3=S 10 amino2=I 9 amino2=T 9 amino4=C 9 amino4=G 9 amino5=G 9 amino5=S 9	amino5=S 17 amino3=I 13 amino4=I 12 amino1=G 11 amino2=S 11 amino2=S 10 amino4=G 10 amino4=S 10 amino5=G 10 amino2=G 9 amino3=N 9 amino4=N 9 amino5=T 9	amino5=G 12 amino4=S 11 amino4=V 11 amino2=V 10 amino3=G 10 amino3=S 10 * amino2=E 9 amino3=V 9 amino4=G 9 *

In NA experiment Table 9, none of 11 rules of H5N6 were same with H6N1, and 2 were same with H10N8. We got the similar results in the experiment by window7 and window 9. In the whole, S(Serine) and G(Glycine) appeared in large numbers, but we didn't regard amino acid in different location as a standard of choosing key factor.

Table 10. Result of nucleoprotein(window5)

	H5N6	H6N1	H10N8
Rules	amino1=R 18 amino4=E 15 amino3=A 12 amino5=G 12 amino1=A 11 amino1=L 11 amino2=G 10 amino3=L 10 amino5=R 10	amino1=R 18 * amino4=E 15 * amino3=A 12* amino5=G 12* amino1=A 11* amino2=S 11 amino1=L 10 amino2=G 10* amino2=R 10 amino3=L 10 * amino5=R 10 *	amino1=R 18 * amino4=E 15 * amino3=A 13 amino5=G 12* amino1=A 11* amino1=I 11 * amino1=L 10 amino2=G 10* amino2=R 10 amino2=S 10 amino3=L 10 * amino5=R 10 *

In NP experiment Table10, 8 of 9 rules of H5N6 were same with H6N1 and H10N8. We got the similar results in the experiment by window7 and window 9. We got the highest sameness rate(88.9% in both) in this experiment. Key factors are R(Arginine) and E(Glutamic acid).

4 Discussion and conclusion

According to Decision tree and Apriori algorithms, we found new-emerging H5N6 has significant similar Matrix protein(M) and Nucleoprotein(NP) sequences compared with H6N1 and H10N8. Therefore, we concluded that H6N1 and H10N8 virus treatment targeting M and NP will be effective to cure H5N6 virus. M2 protein is a channel protein that protons selectively pass through and plays an important role in replication cycle of influenza virus and NP provides important information in viral NA replication [11], [12].

This study provides efficient way to treat avian influenza virus by discovering similar sequences between three kinds of viruses and targeting those sequences with one specific treatment such as M and NP inhibitors. We could find similarities between three influenza viruses. This gives us relevant possibility to make a treatment targeting the inhibitors. If human inhibitors also have relevant similarities, scientists could make an efficient cure for the disease. But we should be cautious to this because if we target the wrong inhibitor due to misunderstanding, it could damage normal genes, which could lead to other serious disorder. For example, a non-replicating cell such as neurons or brain cells, a single-strand break or other type of damage in the transcribed strand of DNA can block RNA polymerase II to transcript gene sequence. Consequently, the body might not be able to make essential proteins, which regulate gene revelation. [13]

Even if the treatment for influenza is devised, there is still remaining concern as mentioned above. Researchers of this disease should notice that making the treatment of the influenza disease is not the only problem, but checking the possibility of malfunction is also important, and that taking care of all is the ultimate goal of public welfare.

References

1. Alexander, Dennis J. "A Review of Avian Influenza in Different Bird Species." *Veterinary Microbiology* 74.1-2 (2000): 3-13. Web
2. *Cumulative Number of Confirmed Human Cases for Avian Influenza A(H5N1) Reported to WHO, 2003-2013* (n.d.): n. pag. Web.
3. Hoffmann, E., J. Stech, I. Leneva, S. Krauss, C. Scholtissek, P. S. Chin, M. Peiris, K. F. Shortridge, and R. G. Webster. "Characterization of the Influenza A Virus Gene Pool in Avian Species in Southern China: Was H6N1 a Derivative or a Precursor of H5N1?" *Journal of Virology* 74.14 (2000): 6309-315. Web.
4. Somsri Wiwanitkit and Viroj Wiwanitkit. "The Three Newest Emerging Influenza H10N8, H5N6 and H6N1: Comparison of Its Morbidity and Mortality." *Asian Pacific Journal of Tropical Biomedicine* 5.5 (2015): 360. Web.
5. Gao, W., A. C. Soloff, X. Lu, A. Montecalvo, D. C. Nguyen, Y. Matsuoka, P. D. Robbins, D. E. Swayne, R. O. Donis, J. M. Katz, S. M. Barratt-Boyes, and A. Gambotto. "Protection of Mice and Poultry from Lethal H5N1 Avian Influenza Virus through Adenovirus-Based Immunization." *Journal of Virology* 80.4 (2006): 1959-964. Web.
6. Pan, Ming, Rongbao Gao, Qiang Lv, Shunhe Huang, Zhonghui Zhou, Lei Yang, Xiaodan Li, Xiang Zhao, Xiaohui Zou, Wenbin Tong, Suling Mao, Shumei Zou, Hong Bo, Xiaoping Zhu, Lei Liu, Heng Yuan, Minghong Zhang, Daqing Wang, Zumao Li, Wei Zhao, Maoli Ma, Yaqiang Li, Tianshu Li, Huiping Yang, Jianan Xu, Lijun Zhou, Xingyu Zhou, Wei Tang, Ying Song, Tao Chen, Tian Bai, Jianfang Zhou, Dayan Wang, Guizhen Wu, Dexin Li, Zijian Feng, George F. Gao, Yu Wang, Shusen He, and Yuelong Shu. "Human Infection with a Novel, Highly Pathogenic Avian Influenza A (H5N6) Virus: Virological and Clinical Findings." *Journal of Infection* 72.1 (2016): 52-59. Web.
7. "Human Infection with Avian Influenza A H6N1 Virus: An Epidemiological Analysis." *National Center for Biotechnology Information*. U.S. National Library of Medicine, n.d. Web. 30 Jan. 2016.
8. Zhang, Tao, Yuhai Bi, Huaiyu Tian, Xiaowen Li, Di Liu, Ying Wu, Tao Jin, Yong Wang, Quanjiao Chen, Ze Chen, Jianyu Chang, George F. Gao, and Bing Xu. "Human Infection with Influenza Virus A(H10N8) from Live Poultry Markets, China, 2014." *Emerg. Infect. Dis. Emerging Infectious Diseases* 20.12 (2014): n. pag. Web.
9. https://en.wikipedia.org/wiki/Decision_tree
10. https://en.wikipedia.org/wiki/Apriori_algorithm
11. http://en.wikipedia.org/wiki/M2_proton_channel
12. http://drugdiscoveryopinion.com/22010/06/targeting-influenza-a-nucleoprotein/
13. Schinecker, T. M. "Human RNA Polymerase II Is Partially Blocked by DNA Adducts Derived from Tumorigenic Benzo[c]phenanthrene Diol Epoxides: Relating Biological Consequences to Conformational Preferences." *Nucleic Acids Research* 31.20 (2003): 6004-015. Web.

MATEC Web of Conferences **69**, 01004 (2016) DOI: 10.1051/matecconf/20166901004
ICCPE 2016

Treatment of various avian influenza virus based on comparison using decision tree algorithm

Jongchan Lee[1], Seojun Kim[2] and Taeseon Yoon[3]

[1]*Natural Science Course, Hankuk Academy of Foreign Studies, Yongin, Republic of Korea*
[2]*Natural Science Course, Hankuk Academy of Foreign Studies, Yongin, Republic of Korea*
[3]*Computer Science and Engineering, Korea University, Seoul, Republic of Korea*

Abstract. Recently, the world is full of anxiety about AI(Avian Influenza). Among Avian Influenza virus subtypes, H5N1 is considered the most threatening to not only birds, but also humans as numerous human cases with high mortality have been reported. Unlike H7N9, which has not been reported infect humans, H5N8 also became infectious to humans due to dramatic mutation. As human infection cases of AI have increased, numerous researchers have been trying to develop an effective treatment against them. Thus, our project group decided to analyze the similarity and difference of H5N1, H5N8, and H7N9, since it would be useful for finding effective treatment of AI, using Decision Tree Algorithm which figures out distinctive factors of given dataset for comparing protein sequences of each viruses. The comparison using Decision Tree Algorithm, which indicates correlation among H5N1, H5N8, and H7N9, will be effective on narrowing the range of attempts on developing treatment for Avian Influenza virus.

1 Introduction

Recently, the world is full of anxiety about AI(Avian Influenza), as many Avian Influenza viruses subtypes have killed not only a number of birds but also other species and even humans. Among them, the highly pathogenic influenza A virus subtype H5N1 is considered the most threatening to both birds and humans.

H5N1 has killed millions of poultry in numerous countries throughout Asia, Europe, and Africa, causing enormous damage. Generally, AI is believed to be infectious only to birds. However, since the first transmission of H5N1 to human was reported in Hong Kong in 1997, 844 cases have been reported until December, 2015 in Cambodia, Egypt, Indonesia, Viet Nam and so on, resulting a high mortality rate of about 50% [1, 2].

Similar to H5N1, H5N8 is an avian influenza virus. As H5N8 devastated South Korea in 2014, about 200 million 3 thousand poultry were killed [3]. It has not been reported to infect humans unlike H7N9, a kind of AI, which became infectious to humans starting from 2013 due to dramatic mutation [4]. However, it seems that H5N8 is fairly capable of infecting humans in that H7N9 went through a dramatic mutation in a relatively short period.

As human infection cases of AI, which has a considerably high mortality rate, have increased, numerous researchers have been trying to develop an effective treatment against H5N1, H7N9. Actually, there are various suggested vaccines that mainly target the glycoprotein of influenza [5].

In this research, we analyzed the similarity and difference among the respective Protein HA, NA, NP, M of H5N1, H5N8, and H7N9 using Decision Tree Algorithm, since it would be useful for finding effective treatment of AI.

2 Materials and methods

2.1 H5N1

H5N1 virus is a subtype of Influenza A virus which triggers highly pathogenic avian influenza (HPAI). One of the most disastrous infections of H5N1 occurred mid-December 2003 to January 2004 [6]. Avian influenza H5N1 has been found in wide areas; its human susceptibility has been reported in South Korea, Japan, Vietnam, Thailand, Cambodia, Indonesia, Pakistan, Laos, and China [7], threatening human health and resulting a plunge in agricultural production [8]. Direct contact with pathogenic materials such as excreta or secretion of poultry seems to cause infection of H5N1 to human. H5N1 has been reported to cause human respiratory problems, fever, pneumonia, and relatively severe cytokine releases [9].

2.2 H5N8

H5N8 virus is a subtype of Influenza A virus. It was first reported from turkey in 1983 Ireland. In January 2014 South Korea, avian influenza that occurred among poultry and caused enormous damage was confirmed as a H5N8 virus [10]. H5N8 viruses cause fever, respiratory problems, and conjunctivitis among poultry.

2.3 H7N9

H7N9 virus is a subtype of Influenza A virus, but was newly revealed as a human infectious virus since the outbreak from China in 2013 [11]. Several H7 viruses such as H7N2, H7N3, and H7N7 had been previously verified as human-infectious. Its infection pathways also include direct and indirect exposure with pathogenic materials. Generally, H7N9 brings deadly pneumonia and fever on human while it causes relatively slight illness among poultry. As a result, it requires researches to distinguish infected poultry from others [12].

2.4 Glycoprotein of Influenza A virus; Hemagglutinin (HA) and Neuraminidase (NA)

Hemagglutinin (HA) and neuraminidase (NA) are among the essential proteins for invasion of virus into host cells; there are 11 type of Influenza A viral proteins – PB1, PB1-F2, PB2, PA, HA, NA, M1, M2, NS1, NEP, and NP. Hemagglutinin is a surface glycoprotein that enables virus to initiate chemical reactions to get into plasmic membranes of infected cell. It combines with monosaccharide sialic acid, accelerating attach on the surface of cell [13]. Neuraminidase is also viral surface glycoprotein that is, on the other hand, responsible for the escape of virus from host cell. It decomposes sialic acid from host cell, facilitating transmission to other cells [14]. The type of hemagglutinin and neuraminidase from virus has been considered as distinguishable elements for classification in influenza viruses. As a result, the name of Influenza A virus reflects the type of HA and NA, resulting H1N1, H5N8, H7N9, etc.

2.5 Decision Tree Algorithm

Decision Tree is a common tool of data mining using tree diagrams. It contains 'nodes' that require decisions for each questions about the relative degree of influence on specific results. Decision Tree Algorithm(ID3 algorithm; Iterative Dichotomiser 3 algorithm) utilizes Decision Tree in order to deal with relevance between given data(x_1, x_2, $x_3, \ldots, x_k$) with major component(Y). Decision Tree algorithm indicates the relationship as

$$(X, Y) = (x_1, x_2, x_3, \ldots, x_k, Y)$$

Decision Tree Algorithm calculates entropy(H) of dataset.

$$H(S) = \sum(-P_i \log(P_i)) \quad (P\text{: given dataset})$$

It extracts rules that show certain dataset which has lowest H; in other words, which has highest relevance with result(Y) sets. This process originates from decision trees judging which data to accept; based on entropy(H) measure [15].

In our research, Decision Tree Algorithm was used for showing interrelation of H5N1, H7N9, and H5N8 viruses in molecular level; amino acidic sequences of each viruses. From extracted rules that indicates the degree of differences among viruses, it is possible to infer distinguishable characteristics of each objects [16]

2.6 Dataset

For the experiments on Influenza A virus; H5N1, H5N8, and H7N9, nucleotide sequences of viral proteins; HA, NA, NP, M derived from NCBI data were used in our research. KF182741.1(HA), KF182743.1(NA), KF182742.1 (NP), KF182744.1(M) are viral protein sequences of H5N1 from Vietnam, KJ476657.1(HA), KJ476659.1(NA), KJ476658.1(NP), KJ476660.1(M) are viral protein sequences of H5N8 from Shanghai, KJ549786.1(HA), KJ549788.1(NA), KJ549789.1(NP), KJ549789.1(M) are viral protein sequences of H7N9 from Shanghai. These sequences of viral protein HA, NA, NP, M were used to compare the corresponding protein sequences of each viruses.

3 Result

We basically extracted data with frequency rates 0.750 or higher in HA. However, we extracted data with frequency rates 0.800 or higher in H7N9 HA under 5 window, 9 window due to observation of relatively excessive rules than other viruses. Also, we extracted data with frequency rates higher than 0.750 in NA.

Table 1. HA Rule Extraction Under 5 Window

virus	Rules	Frequency
H5N1	pos2 = R pos3 = S pos3 = R pos5 = Q	0.750 0.750
H5N8	pos3 = L pos4 = R pos3 = L pos4 = T	0.750 0.750
H7N9	pos3 = D pos4 = L pos3 = T pos4 = E pos3 = M pos1 = L pos3 = F pos3 = S pos5 = G pos2 = G pos3 = N	0.800 0.800 0.800 0.800 0.833 0.800

According to Table 1, It is noticeable that all viruses show their rules extracted. However, rules extracted from H7N9 were distinctively numerous than rules from H5N1 and H5N8. This result is due to relative lack of unique amino acidic features of H5N1 and H5N8 compared to H7N9. It also supports that H7N9 has different amino acid features of HA compared with H5N1, H5N8, and that H5N1 and H5N7 have similarities in some amino acidic

 DOI: 10.1051/matecconf/20166901004

features in HA. Also, it can be assumed that position 3 is the important factors that differentiate each other since position 3 is observed most frequently. Position 1 and 4 also seem to be important factors since they are also observed frequently.

Table 2. HA Rule Extraction Under 7 Window

virus	Rules	Frequency
H5N1	not extracted	
H5N8	pos5 = P	0.750
H7N9	pos5 = H pos1 = D pos 5 = M pos5 = R	0.800 0.750 0.750

Table 3. HA Rule Extraction Under 9 Window

virus	Rules	Frequency
H5N1	not extracted	
H5N8	pos4 = G pos6 = F	0.750
H7N9	pos9 = W pos9 = F pos4 = F pos4 = R pos3 = Q	0.800 0.800 0.800 0.909 0.833

According to Table 2, Table 3, similar to 5 window, rules extracted from H7N9 were distinctively numerous than rules from the H5N1, H5N8. Furthermore, there were no extracted rules of H5N1. This result also supports that H5N1, H5N8 have similar amino acidic features in HA, and H7N9 has distinctive amino acidic features compared to others. In Table 2, it can be assumed that position 5 is the important factors which differentiate each other. In Table 3, position 4 and position 9 are the important factors which differentiate each other. Due to altered window condition, although positions which are considered as important factors changed, the fact that H7N9 has excessive rules compared with H5N1 and H5N8 did not change.

Table 4. NA Rule Extraction Under 5 Window

virus	Rules	Frequency
H5N1	pos3 = P pos4 = D pos3 = T pos4 = D	0.800 0.800
H5N8	pos3 = H	0.800
H7N9	pos3 = G pos4 = A pos1 = A pos4 = I	0.800 0.800

According to Table 4, the results under 5 window, it is noticeable that rules of amino acids at position 3 and position 4 were shared between H5N1, H5N8, and H7N9. Compared to Table 2 and Table 3, NA sequences generally showed equal frequency of extracted rules. It showed identical positions but different corresponding amino acids. We assume that these viruses showed different amino acidic features in NA with each other probably due to the difference in types of viral neuraminidase.

Table 5. NA Rule Extraction Under 7 Window

virus	Rules	Frequency
H5N1	pos2 = P pos5 = S pos1 = G pos3 = V	0.833 0.800
H5N8	pos1 = T pos4 = V pos2 = R pos7 = T	0.800 0.800
H7N9	pos3 = N	0.857

According to Table 5, the results under 7 window, rules of amino acids at position 1, position 2 and position 3 were shown as important factors that differentiates viruses each other. This result that shows existence of relatively equal number of unique rules also supports that H5N1, H5N8, and H7N9 have distinguishable amino acidic features from each other in NA.

Table 6. NA Rule Extraction Under 9 Window

virus	Rules	Frequency
H5N1	pos5 = S pos6 = G pos3 = D pos7 = I	0.800 0.800
H5N8	pos6 = V	0.786
H7N9	pos6 = D pos6 = S pos8 = S pos7 = V	0.833 0.800 0.800

According to Table 6, the results under window 9, we could assume that position 6 works as the important factors for differentiating three viruses because it is observed it is observed most frequently.

Not as our thought, no rule was found in NP and M of H5N1, H5N8, and H7N9. This result is due to the absence of distinguishable amino acidic differences that are huge enough to be considered as key factor. Thus, we could assume that these viruses may indicate similar amino acidic features in NP and M proteins.

4 Conclusion

With the Decision Tree algorithm, we could see the similarities and differences between three typical avian influenza viruses; H5N1, H5N8, and H7N9. Particularly, Decision Tree Algorithm showed that NP and M of H5N1, H5N8, H7N9 have significantly similar amino acidic features. However, NA of three viruses have considerable differences amongst themselves. Also, in HA, H7N9 has distinctively different amino acidic features compared with H5N1 and H5N8, which have similar amino acidic features.

As a result, we could conclude that H5N1 virus treatments targeting NP and M would be effective to H5N8 and H7N9, which also means we can use common treatments to cure all of H5N1, H5N8, and H7N9. Also, we made a conclusion that H5N1 virus treatments targeting HA will be effective to H5N8, not to H7N9,

DOI: 10.1051/matecconf/20166901004

which indicates we can use common treatments to attack both H5N1 and H5N8.

The outbreaks of Influenza A virus have been causing fatal diseases on both human and poultry. Researchers generally search for vaccines and its infection pathways to prevent its spread when new type of Influenza A virus occurs. Thus, if they lacked data of such virus, the treatment would require numerous attempts on finding appropriate viral vaccine. Hopefully, from our research on Influenza A virus (H5N1, H5N8, and H7N9), we investigated the similarity and difference between viruses using data mining Decision Tree Algorithm. As we mentioned above, it suggests the possibility of effective methods for developing treatments on each viruses in terms of time and economy.

Based on its similarity among viruses, we can narrow the range of attempts on developing treatments. Even when a new type of Influenza A virus emerges, it is possible to infer its overall characteristics with indirect bioinformatic research on correlation between known viruses. We hope that our research on Influenza A virus will be widely used for epidemiologic research of various viruses and contribute to finding effective treatment of them.

References

1. Peiris, J. S. M, M. D. De Jong, and Y. Guan. "Avian Influenza Virus (H5N1): A Threat to Human Health." *Clinical Microbiology Reviews 20.2* (2007): 243-67.
2. "Cumulative Number of Confirmed Human Cases of Avian Influenza A(H5N1) Reported to WHO." World Health Organization. N.p., n.d. Web. 17 Jan. 2016.
3. Herriman, Robert. "South Korea H5N8 Avian Flu Update, Virus Spreads to Chicken Farm." *The Global Dispatch RSS*. 26 Jan. 2014.
4. Young, Kim Dae, Kim Hye-Jun, Bae Junhyeok, and Yoon Taeseon. "Examining the Probability of the Critical Mutation of H5N8 by Comparing with H7N9 and H5N1 Using Apriori Algorithm and Support Vector Machine." *IJCTE International Journal of Computer Theory and Engineering 7.2* (2015): 145-48.
5. Jang, S. P., K. H. Park, Y. L. Kim, H. N. Cho, and T. S. Yoon. "Comparison of H5N1, H5N8, and H3N2 Using Decision Tree and Apriori Algorithm." *Journal of Biosciences and Medicines JBM 03.06* (2015): 49-53.
6. "Bird Flu: Global Response Sought," CNN.com, Jan. 27, 2004
7. David P. Fidler, "Global Outbreak of Avian Influenza A (H5N1) and International Law," *American Society of International Law,* January 25, 2004
8. WHO, Avian Influenza A (H5N1) Infection in Humans: Urgent Need to Eliminate the Animal Reservoir-Update 5, Jan. 22, 2004
9. Chan MC, Cheung CY, Chui WH, Tsao SW, Nicholls JM, Chan YO, Chan RW, Long HT, Poon LL, Guan Y, Peiris JS. (2005). "Proinflammatory cytokine responses induced by influenza A (H5N1) viruses in primary human alveolar and bronchial epithelial cells". *Respir. Res.* 6 (1): 135
10. Kang HM, Lee EK, Song BM, Jeong J, Choi JG, et al, "Novel reassortant influenza A(H5N8) viruses among domestic and wild ducks, South Korea, 2014," *Emerg Infect Dis.* 2015 Feb
11. "Frequently Asked Questions on human infection with A(H7N9) avian influenza virus, China," WHO, 2013
12. "Avian Influenza A (H7N9) Virus," Centers for Disease Control and Prevention, 2013
13. White JM, Hoffman LR, Arevalo JH, et al. (1997). "Attachment and entry of influenza virus into host cells. Pivotal roles of hemagglutinin". In Chiu W, Burnett RM, Garcea RL. Structural Biology of Viruses. *Oxford University Press.* pp. 80–104.
14. von Itzstein M (December 2007). "The war against influenza: discovery and development of sialidase inhibitors". *Nature Reviews. Drug Discovery 6* (12): 967–74
15. Go, Eunby, Seungmin Lee, and Taeseon Yoon. "Analysis of Ebolavirus with Decision Tree and Apriori Algorithm." *International Journal of Machine Learning and Computing IJMLC 4.6* (2014): 543-46.
16. Jang, S. P., K. H. Park, Y. L. Kim, H. N. Cho, and T. S. Yoon. "Comparison of H5N1, H5N8, and H3N2 Using Decision Tree and Apriori Algorithm." *Journal of Biosciences and Medicines JBM 03.06* (2015): 49-53.

MATEC Web of Conferences **69**, 01005 (2016) DOI: 10.1051/matecconf/20166901005
ICCPE 2016

Deeper understanding of Flaviviruses including Zika virus by using Apriori Algorithm and Decision Tree

Youjin Yang[1], Bokyung Gu[2] and Taeseon Yoon[3]

[1]*Nature Science Course, Hankuk Academy of Foreign Studies, Yongin, South Korea*
[2]*Nature Science Course, Hankuk Academy of Foreign Studies, Yongin, South Korea*
[3]*Hankuk Academy of Foreign Studies, Yongin, South Korea*

Abstract. Zika virus is spreaded by mosquito. There is high probability of Microcephaly. In 1947, the virus was first found from Uganda, but it has broken out all around world, specially North and south America. So, apriori algorithm and decision tree were used to compare polyprotein sequences of zika virus among other flavivirus; Yellow fever, West Nile virus, Dengue virus, Tick borne encephalitis. By this, dissimilarity and similarity about them were found.

1 Introduction

Today Zika virus is being a very big problem. In Brazil, more than 1.5 million people have become infected since April, 2015. The Zika virus was first found in Uganda in 1947 [1, 2]. As time goes by, zika virus arose in many countries like Malaysia, Micronesia, and Tahiti island in the South Pacific. Now, in recent 2 months, Zika virus has been found in about 39 countries. And, because of zika virus, few people died [3]. Not only Zika virus, but also other viruses (like west nile virus, yellow fever, tick borne encephalitis, and dengue virus) are fatal to people. And all of them belong to flavivirus. Unfortunately, there is no specific treatment of the viruses. Actually, vaccines exist for yellow fever and dengue virus [4, 5] but other viruses (west nile virus, tick borne encephalitis, and zika virus) don't have vaccine. And treatments are only relief from symptoms. This is why we are doing this experiment. An investigation on figuring out similarities and differences between the viruses and yellow fever may lead us to have a deeper understanding about the viruses. By using apriori algorithm and decision tree, we are able to expect to find out the possibility of a treatment that is effective to them.

2 Materials and methods

Materials we used in the experiment are flaviviruses ; Zika Virus, West Nile Virus, Yellow Fever Virus, Tick Borne Encephalitis, and Dengue Virus. Their genome sequences were gathered from the National Center for Biotechnology Information (NCBI). And we use apriori algorithm and decision tree to progress the experiment.

2.1 Zika Virus

In 1947, zika virus was first identified in Uganda rhesus monkey. [1, 2] People can find that the virus is shared to homo-sapiens from mosquito. Conducted three to seven days, people who are infected can notice only a minimal symptoms such as rash, acute fever, conjunctivitis and muscle pain appear. But, roughly 80% cases are an inapparent infection [6]. This virus has high probability of Microcephaly. So many countries warn to woman who can become pregnant [3].

2.2 West Nile Virus

Mainly west nile virus is infected by mosquito, but people can catch it from horse, crow and sparrow. Symptoms like headache, seizure, feeling stiff appear for 2 to 14 days after being infected. The virus can disturb the state of the central nervous system of the brain [7, 8].

2.3 Tick Borne Encephalitis (TBE)

Tick borne encephalitis(TBE) is infected by mite who comes down with the virus. The incubation period for the virus is about 7~14 days. It has symptoms like lack of appetite, vomiting in the incipient stage and the central nervous system like nuchal rigidity or lethargy later [7, 9, 10].

2.4 Yellow Fever

Yellow Fever is like hemorrhagic fever that is fashionable disease in Africa and North America. The arbovirus that causes yellow fever is transmitted by mosquitoes [11]. After the incubation period of three to six days, people can experience symptoms like fever, cold fit, headache, and vomiting [12].

2.5 Dengue Virus

Dengue virus exists in saliva of Aedes albopictus. When the mosquito the blood of people or animals the virus is injected into a person and then the disease goes to the person. The disease occurs 5~7 days after being bitten by mosquitoes. Fever and skin rash are the major symptoms [13-15].

2.6 Apriori algorithm

The Apriori Algorithm is the one of Algorithms used for data mining [14, 16]. In this research, We used apriori algorithm to find frequency of polyprotein sequence of Flavivurses. We divided 3 window rules; 9-window, 13-window and 17-window. The number of window represents is the number of sequences of virus which we have disposed before applying algorithm. We can compare with Zika virus and other Flaviruses to use frequency of polyprotein sequence. In other words, we can find similarity or dissimilarity about Zika virus and the viruses.; Dengue virus, West nile virus, Yellow fever, and Tick borne encephalitis.

2.7 Decision tree

One of the common data mining methods is decision tree. It uses branching method by drawing leaves and branches to generate model which expects target variable based on several input variables [6, 16]. Decision tree is a graph that every internal node corresponds to input variables and each branch corresponds to possible outcome of the input variables. And leaf node is a value of target variable when each input node has a level of route from root node to leaf node. Decision tree stretches continuously by adding the input variables. It expands until node of subset equals to target variable or the new predictive value cannot be added caused by division. If we use decision tree, it is effective to find out differences between data [13, 17]. So in our experiment, decision tree method is applied in order to compare and contrast the viruses.

3 Result

3.1 Apriori algorithm

Using apirori algorithm, we found out results like these.

Zika Virus
1. pos3 is G 47
2. pos8 is L 47
3. pos1 is G 41
4. pos5 is G 41
5. pos7 is L 40
6. pos2 is L 38
7. pos9 is L 38

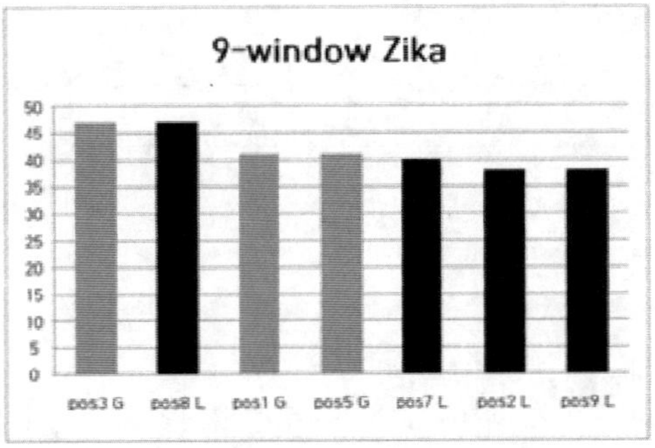

Figure 1. 9-window Zika polyprotein sequence.

Fig. 1 is the result we found out by analyzing Zika virus using apriori algorithm window-9. Results from other viruses are appeared in similar patterns. Averagely, about 40 rules were found in each virus (including 9, 13 and 17 windows), so we got total 201 rules. We analyze rules of polyprotein sequences by the appearance frequency. And then, the most symbolic rule of each virus was chosen and was used for finding similarity between 5 viruses. Tables below are the results.

Table 1. Apriori 9-window rules

Type	rule
Dengue virus	pos13=L 29
TBE	pos17=G 27
Yellow fever	pos10=L 27
West nile virus	pos1=L 26 pos13=A 26 pos15=V 26
Zika virus	pos1=G 30

Table 2. Apriori 13-window rules

Type	rule
Dengue virus	pos1=L 42
TBE	pos2=G 44 pos3=L 44
Yellow fever	pos5=L 43
West nile virus	pos7=L 46
Zika virus	pos3=G 47 pos8=L 47

Table 3. Apriori 17-window rules

class (virus)	Dengue virus	TBE	West Nile	Yellow Fever	Zika Virus
Dengue virus	86	78	92	67	54
TBE	81	64	89	82	64
West Nile	96	68	86	70	62
Yellow Fever	111	76	76	70	46
Zika virus	99	87	87	68	40

Table 1, Table 2, Table 3 show an analysis of amino acid sequences of 5 flaviviruses in 9, 13, 17 window. 9-window: Dengue virus amino1 Leucine, Tick borne encephalitis amino2 Glutamine and amino3 Leucine, Yellow Fever animo5 Leucine, West Nile amino7 Leucine, Zika Virus amino3 Glutamine and amino8 Leucine. 13-window: Dangue Virus amino1 Leucine and amino7 Valine, Tick borne encephalitis amino8 Leucine, Yellow Fever amino12 Glutamine, West Nile amino2 Glutamine and amino9 Leucine, Zika Virus amino7 and 13 Leucine. 17-window: Dengue Virus13 Leucine, Tick borne encephalitis amino17 Glutamine, Yellow Fever amino10 Leucine, West Nile amino1 Leucine and amino13 Alanine and amino15 Valine, Zika Virus amino1 Glutamine.

3.2 Decision tree

Table 4. Decision tree 9-window rules

Type	**rule**
Dengue virus	pos13=L 29
TBE	pos17=G 27
Yellow fever	pos10=L 27
West nile virus	pos1=L 26 pos13=A 26 pos15=L 26
Zika virus	pos1=G 30

Analyzing Table 4, we found out that Dengue virus is similar with other virus. On the other hand, Zika virus is different with other viruses. Dengue virus, tick borne encephalitis, west nile virus, and yellow fever virus have their own characteristics. But, lack characteristics was found at Zika virus.

Especially in Table 6, west nile virus is similar with dengue virus. Except this, similarity of other 4 viruses is high. And none of their own characteristics of all 5 flaviviruses was found.

Compared with Table 4, the number of rules of zika virus got higher at Table 5. Zika virus is similar with other viruses and has its own characteristics like dengue virus and tick borne encephalitis. In contrast, viruses such as west nile and yellow fever are different with other viruses and have less characteristics of their own.

Table 5. Decision tree 13-window rules

class (virus)	Dengue virus	TBE	West Nile	Yellow Fever	Zika Virus
Dengue virus	59	53	42	42	65
TBE	52	67	43	35	66
West Nile	62	58	39	41	65
Yellow Fever	60	51	49	36	67
Zika virus	54	50	39	50	71

Table 6. Decision tree 17-window rules

class (virus)	Dengue virus	TBE	West Nile	Yellow Fever	Zika Virus
Dengue virus	38	37	60	35	30
TBE	37	35	42	45	42
West Nile	42	41	36	40	43
Yellow Fever	30	45	45	36	45
Zika virus	36	50	40	35	41

4 Discussion and conclusion

We measured the frequency of 5 flaviviruses' polyprotein sequences using apriori algorithm. Experiments were separated into 3(9-window, 13-window, 17-window) and were done on each flavivirus. We can find out that G(glutamine) and L(leucine) showed higher frequency of all viruses. Exceptively V(valine) was frequently found in 13-window of dengue virus, V(valine) and A(alanine) were frequently found in 17-window of West Nile virus. By comparing and contrasting zika virus with other 4 viruses, we expected that we can find out treatment of zika virus.

To know exact correlation between zika virus and other 4 viruses, we used decision tree set to 10-fold cross validation. Type of proteins that showed higher frequency is almost same at apriori. However, we can find out each virus has its own characteristics. Namely, there may be some difficulties at combining 4 flaviviruses to develop treatment of zika virus.

So we think that further research have to be progressed.

References

1. Dick, G.w.s, S.f Kitchen, and A.j Haddow. Transactions of the Royal Society of Tropical Medicine and Hygiene 46.5 **209-20** (1952)
2. Dick, G.w.a. Transactions of the Royal Society of Tropical Medicine and Hygiene 46.5 **521-34** (1952)
3. M. Robert W., J. Homan, M. V. Callahan, J. Glasspool-Malone, L. Damodaran, A. D. B. Schneider, R. Zimler, J. Talton, R. R. Cobb, I. Ruzic, J. Smith-Gagen, D. Janies, and J. Wilson. PLoS Negl Trop Dis PLOS Neglected Tropical Diseases 10.3 (2016)
4. Mcarthur, Monica A., Marcelo B. Sztein, and Robert Edelman. "Dengue Vaccines: Recent Developments, Ongoing Challenges and Current Candidates." Expert Review of Vaccines 12.8 **933-53** (2013)
5. M. Milton, F. Da Silva Pereira Cruz, M. T. Cordeiro, M. A. Da Motta, K. M. De Melo Cassemiro, R. De Cassia Carvalho Maia, R. C. Bressan Queiroz De Figueiredo, R. Galler, M. Da Silva Freire, J. T. August, Ernesto T. A. Marques, and Rafael Dhalia. PLoS Negl Trop Dis PLOS Neglected Tropical Diseases 9.4 (2015)
6. Hayes, Edward B. Emerg. Infect. Dis. Emerging Infectious Diseases 15.9 **1347-350** (2009)
7. B. Cécile, M. Jimenez-Clavero, A. Leblond, B. Durand, N. Nowotny, I. Leparc-Goffart, S. Zientara, E. Jourdain, and S. Lecollinet. International Journal of Environmental Research and Public Health IJERPH 10.11 **6039-083** (2013)
8. Winkelmann, R. Evandro, H. Luo, and T. Wang. F1000Research F1000Res (2016)
9. M. Ecker, S. L. Allison, T. Meixner, and F. X. Heinz. Journal of General Virology 80.1 **179-85** (1999)
10. P. Vasiliki, S. Geroy, E. Diza, A. Antoniadis, and A. Papa. Vector-Borne and Zoonotic Diseases 7.4 **611-16** (2007)
11. B. Matthias, D.A. Groneberg, D. Klingelhoefer, and A. Gerver. Parasites & Vector Parasit Vectors 6.1 **331** (2013)
12. Gardner, L. Christina, and K. D. Ryman. Clinics in Laboratory Medicine 30.1 **237-60**(2010)
13. E.B. Go, S.M. Lee, and T.S. Yoon. International Journal of Machine Learning and Computing IJMLC **543-46** (2014)
14. H.S. Kim, J.Y. Yoo, and T.S. Yoon. International Journal of Computer and Communication Engineering IJCCE 294-301(2015)
15. R. Rodriguez-Roche, and Ernest A. Gould. BioMed Research International 2013 **1-20** (2013)
16. Y. S. Kim, S. M. Kim, J. W. Lee, J. Ann , J. Lim, and T. S. Yoon. Intelligent Computing Theories and Methodologies Lecture Notes in Computer Science **426-35** (2015)
17. Jang, S. P., K. H. Park, Y. L. Kim, H. N. Cho, and T. S. Yoon. Journal of Biosciences and Medicines JBM 03.06 **49-53**(2015)

Data mining and analysis of bacillus virus

Jiwoo An[1], ChanWoo Kim[1], Seohyun Moon[1], Jungeun Huh[1] and Taeseon Yoon[2]

[1]*Natural Science, Hankuk Academy of foreign studies, Yong-in, Korea*
[2]*Computer Science, Korea University, Seoul, Republic of Korea*

Abstract. Anthrax- can be found naturally in soil and commonly affects domestic and wild animals around the world. Bacillus anthracis is mostly common viruses of Anthrax. There are some more occurred in similar DNA sequences: 3 Bacillus Viruses: Bacillus anthracis, Bacillus cereus, and Bacillus thuringiensis. This is a report about analyzing the similarities between the Bacillus viruses by investigating the frequency of amino acid and finding the difference between those three viruses based on the gene. Those diseases are infected by parasite and host animals, and cause muscle pain. Therefore, we can conclude that Leucine is a protein that plays a significant role in causing muscle pain. Secondly, In Analysis of decision tree, there are only little differences between each classes. The classes represent positions that include representative protein. The very position that windows mention is their difference.

1 Introduction

Bacillus, a bacteria, is a bar and cylindrical shaped aerobacter. There are a 34 species of bacillus virus, such as bacillus cereus, which causes food decay and has flagellum but doesn't have any capsule (different from bacillus anthracis), and bacillus thuringiensis, which produce sterilized toxin. However, most of them are avirulence, but bacillus anthracis is the only virus that causes anthrax. Therefore, Bacillus anthracis was utilized for a number of biochemical weapons. Since each bacillus virus causes different influence to the nature, we attempt to reveal the difference between bacillus cereus, bacillus thuringiensis and bacillus anthracis. We use decision tree program and apriori program as classification standards. Apriori is used to measure the frequency while the decision tree classifies elements that are unique.

2 Apriori experiment

We did an experiment about three bacillus viruses: bacillus anthracis, bacillus cereus, and bacillus thuringiensis. We made the results using apriori algorithm and decision tree algorithm. Before having an apriori experiment, we divided three bacillus into three classes: 13window, 17window, and 19window. Since the lengths of each bacillus are different, we divided them into different windows according to their lengths that the comparison with the viruses' position similar is possible when the sequences are repeated. We got similar consequences from each cases, thus overall we could conclude that acid I-Leucine has the highest frequency and that all three cases are involved in the actions occurring in the body.

2.1 Captions/numbering

We use apriori algorithm to figure out which protein has highest frequency in each of three bacillus viruses. Apriori algorithm is used to find out the most frequent protein, like frequent item set mining algorithm, and association rule. learning over transactional databases. So, we use this algorithm to know which amino acid mostly consists bacillus-Leucine-, why Leucine shows the highest frequency and how Leucine affects bacillus and our bodies.

2.2 Apriori results

We got exactly same results in all experiment-all windows of three viruses: Leucine.

Table 1. All experiment-all windows of three viruses

virus/ windows	Bacillus anthracis	Bacillus cereus	Bacillus thuringiensis
13 windows	I-Leucine (13850)	I-Leucine (13989)	I-Leucine (13925)
17 windows	I-Leucine (10591)	I-Leucine (10697)	I-Leucine (10649)
19 windows	I-Leucine (9476)	I-Leucine (9571)	I-Leucine (9528)

Numbers of Table 1. represent the average figure of Leucine found in each Amino groups.

DOI: 10.1051/matecconf/20166901006

Following Tables 2, 3, 4 are each about frequency analysis about Leucine in bacillis anthracis, bacillus cereus, and bacillus thuringiensis at 19 windows.

Table 2. Analysis of bacillus anthracis

position	window		
	13 (count)	*17 (count)*	*19(count)*
Position 1	L(13815)	L(10542)	L(9541)
position 2	L(13798)	L(10586)	L(9609)
position 3	L(13908)	L(10585)	L(9372)
position 4	L(13930)	L(10391)	L(9345)
position 5	L(13813)	L(10813)	L(9302)
position 6	L(13849)	L(10469)	L(9538)
position 7	L(13837)	L(10649)	L(9490)
position 8	L(13967)	L(10600)	L(9433)
position 9	L(13943)	L(10635)	L(9590)
position 10	L(13897)	L(10509)	L(9514)
position 11	L(13679)	L(10633)	L(9515)
position 12	L(13763)	L(10714)	L(9488)
position 13	L(13849)	L(10533)	L(9524)
position 14		L(10509)	L(9486)
position 15		L(10683)	L(9482)
position 16		L(10516)	L(9382)
position 17		L(10672)	L(9390)
position 18			L(9636)
position 19			L(9402)

Table 3. Analysis of bacillus cereus

position	window		
	13 (count)	*17 (count)*	*19(count)*
position1	L(14030)	L(10713)	L(9519)
position 2	L(13937)	L(10828)	L(9576)
position 3	L(14040)	L(10537)	L(9715)
position 4	L(13983)	L(10700)	L(9538)
position 5	L(13906)	L(10688)	L(9541)
position 6	L(13975)	L(10625)	L(9586)
position 7	L(13955)	L(10630)	L(9696)
position 8	L(14077)	L(10782)	L(9592)
position 9	L(13943)	L(10679)	L(9523)
position 10	L(14207)	L(10612)	L(9546)
position 11	L(13885)	L(10753)	L(9569)
position 12	L(14004)	L(10889)	L(9561)
position 13	L(13914)	L(10841)	L(9574)
position 14		L(10623)	L(9719)
position 15		L(10668)	L(9387)
position 16		L(10535)	L(9685)
position 17		L(10753)	L(9548)
position 18			L(9471)
position 19			L(9510)

2.2.1 Results analysis

As you see in the tables ahead, results of all three bacillus is same-Leucine. Even though there are small differences among figures-exactly show how much time Leucine had been detected-, such things are so insignificant that it cannot cause substantive distinction inside the organism. Therefore, we could figure out that Leucine plays important role in most bacillus virus. For instances, most effects of amino acids on bacillus signaling are abolished by lowering the concentration of Leucine.

Table 4. Analysis of bacillus thuringiensis

position	window		
	13 (count)	*17 (count)*	*19(count)*
position1	L(13967)	L(10492)	L(9652)
position 2	L(14027)	L(10554)	L(9365)
position 3	L(13942)	L(10753)	L(9570)
position 4	L(13919)	L(10592)	L(9589)
position 5	L(13869)	L(10743)	L(9476)
position 6	L(14106)	L(10772)	L(9361)
position 7	L(13750)	L(10679)	L(9480)
position 8	L(13904)	L(10676)	L(9452)
position 9	L(13877)	L(10667)	L(9455)
position 10	L(13976)	L(10616)	L(9595)
position 11	L(13870)	L(10597)	L(9647)
position 12	L(13925)	L(10696)	L(9500)
position 13	L(13898)	L(10649)	L(9624)
position 14		L(10613)	L(9652)
position 15		L(10657)	L(9468)
position 16		L(10708)	L(9474)
position 17		L(10566)	L(9505)
position 18			L(9607)
position 19			L(9558)

2.2.2 Leucine and myalgia (How Leucine effects on our bodies)

Bacillus anthracis, which is infectious to animals, is endemic(Anthrax)-a disease which regularly transmit infection to regional societies-. In some cases, it is infected by physical contacts, thus mostly it occurs in certain region or tribes. You would suffer from myalgia, serious cough, dyspnea, papule, and so on. Not only anthrax but also generous endemics-such as malaria, dengue fever, yellow fever, typhoid fever, cholera, etc-contain those symptoms, especially myalgia. And apriori experiment results of those viruses are the same: Leucine. Therefore, we may concluded that Leucine consists most of endemic viruses and brings out myalgia.

The experiment is adapting Quinla's C 5.0 [3] algorithm and rule extraction method. See 5.0 program is used for the experiment.

Fig. 1 shows abstract shape of common dicision tree, which we used, and Fig. 2 shows the part of decision tree and condition to get to the next node at particular situation.

3 Decision tree

3.1 Decision tree algorithm

We use Decision tree algorithm to find out the difference between three bacillus viruses. Decision tree algorithm showed position of amino acid. So, we used this algorithm to know how can we classify best according to position. Thus, we utilized this algorithm in order to figure out how we can classify the best regarding to position.

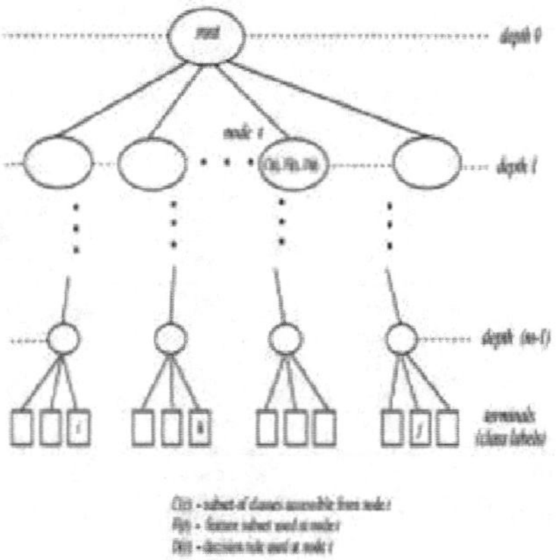

Figure 1. manual form of decision tree [1]

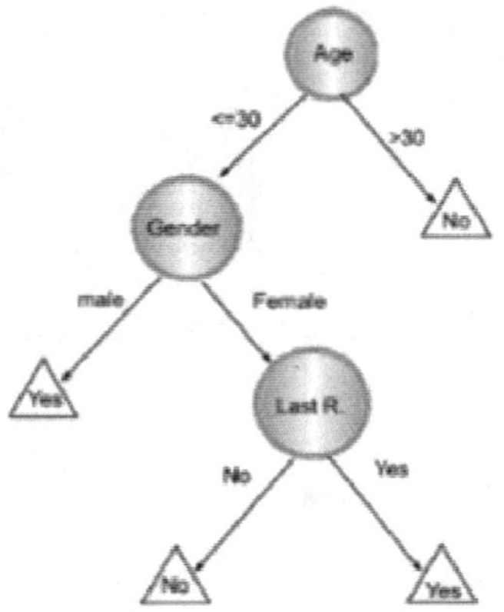

Figure 2. Decision tree presenting response to direct mailing.[2]

3.2 Decision tree

There are mainly two types of decision trees according to usage. Classification tree which is used for classification tasks and regression tree, used for regression tasks. [4]

Our usage of the decision tree is the former one, to classify a case by starting from the tree root and moving it till a leaf is encountered. The case's result for the test at individual non-leaf decision node is decided and focus moves to the subtree's root, matching to this outcome. After this process ultimately (and naturally) generates to a leaf, it is expected that the class of the case to be that noted at the leaf. [3]

A path is made between the root of the decision tree and one of its leaves when the tree is used to classify a case. To reach a single leaf, it must accord with entire conditions on the path. We can know that because of this, decision tree is highly related to rule induction [5]

3.3 Decision tree results

According to the Table 1, we can see that those 3 viruses have a big similarity. Which means, it is hard to find out specificity at each virus. By this result, we can assume that all of these viruses are derived from one same virus and there are small proportion of mutein. According to Table 5, we have to notice rule extraction under 13 window showed their rules with amino acid at position 11. Glutamic Acid (E), Glysin(G), are Isoleucine(I) are all presented on 11th position in 13 window, and this common position represents the property of bacillus virus. Moreover, second position is also playing an important role, presenting G on class 1, N on class 2, and G and C on class 3. In Table 6, we have to notice that under 17 window, rules were extracted at position 16. At Table 7, we can find out under 19 window, position 8 was the place where rules were extracted. Also, we can see that those position played important role differentiating each subtype from each other.

Following Tables 5, 6, 7 are rule extraction due to the position of amino acid.

Table 5. 13 windows

Virus	Rule		
Class 1	position2=G	position10=I	position11=E
	position2=L	position11=G	
	position3=V	position11=I	
	position4=I	position11=F	
	position6=W	position11=L	
Class 2	position2=N	position10=P	position11=E
	position2=A	position11=G	
	position3=H	position11=I	
	position4=A	position11=F	
Class3	position2=G	position10=V	position11=E
	position2=C	position11=G	
	position3=S	position11=I	
	position4=M	position11=F	
	position6=E	position11=L	

Table 6. 17 windows

Virus	Rule	
Class 1	position8=R	position16=I
	position7=S	position8=C
	position4=G	position8=A
	position8=W	position14=E
Class 2	position8=R	position16=S
	position7=K	position8=C
	position4=E	position8=A
	position8=W	position14=L
Class 3	position8=R	position16=Q

	position7=H	position8=C
	position8=W	position14=F

Table 7. 19 windows

Virus	Rule	
Class 1	position4=H	position16=A
	position5=Q	position16=E
	position1=M	position16=V
	position10=Y	position16=N
Class 2	position4=Q	position16=A
	position5=S	position5=E
	position1=G	position16=V
	position10=Q	position16=N
Class 3	position1=H	position16=V
	position4=C	position16=A

4 Conclusion

Firstly, In Analysis of apriori, Three kinds of bacillus virus have Leucine the most (among.....). Leucine is a virus detected in endemic diseases such as malaria [6], dengue fever [7], typhoid [8], serious cough, dyspnea, papule and so on [9,10]. Those diseases are infected by parasite and host animals, and cause muscle pain. Therefore, we can conclude that Leucine is a protein that plays a significant role in causing muscle pain. Secondly, In Analysis of decision tree, there are only little differences between each classes. The classes represent positions that include representative protein. The very position that windows mention is their difference. We are looking forward to conceive natural treatment(not the artificial one) to cure endemic disease. Using common property of bacillus virus that it has Leucine, which is contained in endemic disease, we are planning to analyze the similarity and develop the natural treatment.

References

1. Safavian, S. Rasoul, and David Landgrebe. "A survey of decision tree classifier methodology." *IEEE transactions on systems, man, and cybernetics* **21.3** (1991): 660-674.
2. Rokach, Lior, and Oded Maimon. "Top-down induction of decision trees classifiers-a survey." *Systems, Man, and Cybernetics*, Part C: Applications and Reviews, IEEE Transactions on **35.4** (2005): 476-487.
3. Quinlan, J. Ross. C4. 5: programs for machine learning. Elsevier, 2014.
4. L. Rokach and O. Maimon, Data Mining with Decision Trees: *Theory and Applications, World Scientific Pub Co Inc.*, (2008).
5. Quinlan, J. Ross. "Simplifying decision trees." *International journal of man-machine studies* **27.3** (1987): 221-234.
6. Lee, Jung-Yub, Su-Min Song, Ji-Woong Seok, Bijay Kumar Jha, Eun-Taek Han, Hyun-Ouk Song, Hak-Sun Yu, Yeonchul Hong, Hyun-Hee Kong, and Dong-Il Chung. "M17 Leucine Aminopeptidase of the Human Malaria Parasite Plasmodium Vivax." *Molecular and Biochemical Parasitology* **170.1** (2010): 45-48.
7. Fatima, Zareen, Muhammad Idrees, Mohammad A. Bajwa, Zarfishan Tahir, Obaid Ullah, Muhammad Q. Zia, Abrar Hussain, Madiha Akram, Bushra Khubaib, Samia Afzal, Saira Munir, Sana Saleem, Bisma Rauff, Sadaf Badar, Mahrukh Naudhani, Sadia Butt, Mahwish Aftab, Liaqat Ali, and Muhammad Ali. "Serotype and Genotype Analysis of Dengue Virus by Sequencing Followed by Phylogenetic Analysis Using Samples from Three Mini Outbreaks-2007-2009 in Pakistan." *BMC Microbiology BMC Microbiol* **11.1** (2011): 200.
8. Kariuki, Samuel, Gunturu Revathi, John Kiiru, Doris M. Mengo, Joyce Mwituria, Jane Muyodi, Agnes Munyalo, Yik Y. Teo, Kathryn E. Holt, Robert A. Kingsley, and Gordon Dougan. "Typhoid in Kenya Is Associated with a Dominant Multidrug-Resistant Salmonella enterica Serovar Typhi Haplotype That Is Also Widespread in Southeast Asia." *Journal of Clinical Microbiology. American Society for Microbiology (ASM)*, n.d. Web. 29 Sept. (2015).
9. "Cholera: Causes, Symptoms, Treatment, and Prevention." *WebMD. WebMD*, n.d. Web. 29 Sept. (2015).
10. "Anthrax Causes, Symptoms, Treatment - Anthrax Signs and Symptoms - EMedicineHealth." *EMedicineHealth. N.p.*, n.d. Web. 29 Sept. (2015).

MATEC Web of Conferences **69**, 01007 (2016) DOI: 10.1051/matecconf/20166901007
ICCPE 2016

Comparison of CIV, SIV and AIV using Decision Tree and SVM

Hyorin Park[1], Yoojin Park[1], Yerin Moon[1] and Taeseon Yoon[2]

[1]*Natural Science, Hankuk Academy of Foreign Studies, HAFS, Yongin-si, Republic of Korea*
[2] *Department of Computer Science and Engineering, Korea University, KU, Seoul, Republic of Korea*

Abstract. The H3N2, the canine influenza virus has numerous types of animal hosts that can live and reproduce on. They mostly settle on pigs and birds. However, some concerned voices are rising that there is high possibility that humans could be an additional victim for the canine flu. Consequently, our project group expect that the information about the H3N2's DNA are valuable, since the information could attribute to development of vaccine and medicine. In the experiments of analysing the properties of CIV, Canine Influenza Virus with the comparison of SIV, Swine Influenza Virus and AIV, Avian Influenza Virus with the decision tree and SVM, Support Vector Machine. The result came out that CIV, SIV and AIV are alike but also different in some aspects.

1 Introduction

Recently, various types of influenzas have broken out worldwide. The influenzas can be thought of as similar as each other, but can also be thought as the origins of separate diseases. Determining the ranking of them with their impact to human, to seek the effective treatment, maybe the three-AIV, SIV, and CIV- will win the most of all the influenzas. In our experiment, the sample of viruses was composed of Nev, Nef, Tat and Vif each originating from human, swine and avian influenza virus. The assortments of the isolate implies that gene classification exists. [1]

Specific investigation on gene assortments and the 8 particular gene segments and DNA sequences among the H3N2/AIV/SIV/CIV is the core of the report in terms of finding the treatment of H3N2 CIV. The remedy of H3N2 AIV and SIV could be substituted to potential cure of H3N2 CIV. 'SVM' and 'decision tree' algorithm would be used in order to analogue the resulting DNA sequences and similarities of the 3 viruses. The common gene of the 3 viruses includes the fact that swine, avian, human cells are the possible host or incubator of the virus and the restriction of virus from being cloned could result the destruction of the virus's ability.

2 Background knowledge

2.1 Virus

2.1.1 Swine Influenza virus

The H3N2 Swine Influenza Virus(SIV) originates from the H1N1 Influenza A virus. It produces fever, lethargy, sneezing, coughing, difficulty breathing and decreased appetite in pigs. [2] In humans, on the other hand, the symptoms of the 2009 "swine flu" H1N1 virus are similar to those of influenza-like illness in general. Symptoms include fever, cough, sore throat, body aches, headache, chills and fatigue. [3] The 2009 outbreak has also shown an increased percentage of patients reporting diarrhoea and vomiting.

2.1.2 Canine Influenza virus

Canine Influenza Virus(CIV) is well known as one of the most highly pathogenic subtype of the influenza A virus. Symptoms vary from mild to harsh ones, including a cough which lasts for approximately 30 days, possibly a nasal discharge, high fever and - in extreme cases – pneumonia. [4] The vaccine for this virus has been developed in 2009, but the effectiveness or safety has not yet been verified. [5]

2.1.3 Avian Influenza virus

Avian Influenza Virus (AIV) is an influenza caused by viruses adapted to birds, and it is traced back to its original form, the H5N1 AIV, which is noted for its fatalness. Symptoms include fever, cough, sore throat, body aches, headache, chills and fatigue. [6]

2.2 Algorithm

2.2.1 Decision Tree

A decision tree is a tree-like graph or model composed of node and branches. The decision support tool are openly utilized statistical analysis for classification of category type inputs. [7] Several internal nodes contain questions associated with data items, branches down the original source into subsets. Following branches represents the outcome of test and leaf nodes are directed to a class node. [8]

2.2.2 Support vector machine

SVM(Support Vector Machine) is supervised learning model for data analyzing algorithms and it is used to classification and regression analysis. When a group of data is given, SVM training algorithm builds a model that decide which category to involve the data, based on existing data. SVM model is represented as a widest gap, border of separated categories, in space. [9] New examples are then mapped into that same space and predicted to belong to a category based on which side of the gap they fall on. SVM is used by not only linear classifications but also non-linear classification. For SVM to perform a non-linear classification efficiently, kernel trick is needed to implicitly map their inputs into high-dimensional feature spaces.

2.3 Protein

2.3.1 Nef protein

Nef(Negative Regulatory Factor) protein is a 27-35kDa protein. The protein is encoded by primate lentiviruses which include AIV-1, AIV-2 and SIV. Nef promotes T-cell activation and the establishment of two basic attributes of AIV infections. Nef regulates the cell surface expression of CD4 and Lck. [10] Enrichment of active Lck induces the production of Interleukin 2 which activates the growth, differentiation and proliferation of T-cells. Differentiated T-cells creates a new population of cells in which AIV-1 could infect. In short, Nef protein stands for manipulation of the host's cellular machinery AIV-1 replication in terms of "Negative Factor."

2.3.2 Env protein

Env(Envelop Protein) protein is a protein that forms the viral envelope. The expression of the env gene induces retroviruses to attach to specific cell target and to infiltrate the cell membrane. The env gene codes for the gp160 protein, which is cleaved into gp120 and gp41 by Furin. The glycoprotein gp120 bins to CD4 receptor on the target cell with a receptor including the helper T-cell. Replication cycle of AIV is related with the env gene in which gp120 has been the subject of AIV vaccine research while CD4 receptor binding is an important step in AIV infection. In addition, glycoprotein gp41, which bound to gp120, enables AIV to enter the cell in another step of AIV infection.

2.3.3 Tat protein

In molecular biology, Tat(Trans-Activator of Transcription) is a protein that is encoded for by the tat gene in AIV-1. Tat is a regulatory protein that drastically enhances the efficiency of viral transcription from the LTR promoter and replication. Tat has an unusual property for a transcription factor. It can be released and enter cells freely, yet still retain its activity, enabling it to up-regulate a number of genes.

2.3.4. Vif protein

Vif(Viral infectivity factor) is a protein found in AIV and other retroviruses and is essential for viral replication. It have human enzyme APOBEC(cytidine deaminase enzyme that mutates viral nucleic acids) cause hypermutation of the viral genome, rendering it dead-on-arrival at the next host cell by ubiquitination and cellular degradation . [11] As result, it disrupts antiviral activity. Thus APOBEC plays key role to defend retroviral which AIV-1 has overcome by the acquisition of vif. [12]

3 Method

3.1 Decision Tree

We used decision tree to analyse the overlapped rule among AIV, SIV, and CIV. Dividing the DNA sequence in 10 parts, we tested the particular protein sequence of env, nef, tat, and vif in 5,7,9 window using 10-fold cross validation. Sequences with confidence above 75% were only considered as 'valid'.

3.2 SVM

We used RBF, Polynomial function, normal function, and sigmoid function in this experiment. The X-axis of graph is the kind of function we used, and Y-axis is accuracy. We progressed the experiment in each windows and calculated the average. To test SVM's validity, we compare Accuracy rate. At a higher accuracy, more similarity. [13]

4 Result

4.1 Decision Tree

Decision Tree was used to analyse common rules between DNA sequences of AIV, SIV and CIV. The DNA sequences are divided into 10 subsets and the experiment was held in 10 fold cross validation for each 5 window, 7 window and 9 window rules of specific proteins: ENV, NEF, TAT and VIF. Under all the extracted rules, only those with frequency of over 0.75 were considered as valuable. [14] In the experiment, AIV refers to class 1, SIV refers to class 2 and class 3 represents CIV. In interpreting the result, a particular amino acid has a tendency of being repeatedly observed in a certain position. In the protein ENV, AIV seemed to have the most powerful impact among the three viruses in classification. In protein NEF and TAT, CIV was dominant under rules. On the other hand, SIV and CIV were the main subjects of classification, so the influence of AIV was small in protein VIF.[15]

Table 1. Rule extraction under 5 window of ENV.

Class	Rule	Frequency
class1	pos 1=G pos 4=M	0.750
class2	pos 1=G pos 4=V	0.800
class3	pos 1=L pos 2=R	0.800

In Table 1, the result represents that AIV has considerably mixed with SIV and CIV. In addition, class 2 and class 3 overlapped common sequences with other classes, so the DNA sequences between three viruses are quite evenly mixed.

Table 2. Rule extraction under 7 window of ENV.

Class	Rule	Frequency
class1	pos 2=A pos 4=F	0.750
class2	pos 4=D pos 5=T	0.750
class3	pos 2=S pos 4=S	0.857/0.833

In Table 2, AIV showed high tendency to follow the rules of class 1 while SIV almost equally combined with other classes. On the other hand, CIV refer to class 2 in relatively high percentage.

Table 3. Rule extraction under 9 window of ENV.

Class	Rule	Frequency
class1	pos 7=Q pos 9=H	0.800
class2	pos 7=Q pos 9=R	0.800
class3	pos 7=Q pos 9=N	0.800

In Table 3, AIV, SIV and CIV have the common amino acid Q in position 7 and the three viruses are mixed equally.

Table 4. Rule extraction under 5 window of NEF.

class	Rule	Frequency
class1	pos 3=V pos 5=Q pos 4=H _	0.750 0.750
class2	pos 3=V pos5=L pos 3=M _	0.750 0.800
class3	pos 3=V pos5=V	0.750

In Table 4, we observed the considerably combined DNA sequences among AIV, SIV and CIV.

Table 5. Rule extraction under 7 window of NEF.

class	Rule	Frequency
class1	pos 1=T _	0.750
class2	pos 1=A pos7=G	0.750
class3	_	_

In Table 5, under certain combining of the three viruses, AIV had high tendency to class 3 and SIV was likely to refer to class 2 in highest percentage.

Table 6. Rule extraction under 9 window of NEF.

class	Rule	Frequency
class1	pos 2=C _	0.750

In Table 6, largest number of appearances were observed in pos 2=C of AIV, pos 2=F of SIV and pos 9=N in CIV. The percentile showed that the three viruses are evenly mixed in protein NEF.

Table 7. Rule extraction under 5 window of TAT.

class	Rule	Frequency
class1	pos1=C pos5=P pos1=H	0.83
class2	pos1=D pos5=P	0.75
class3	pos1=E pos5=P	0.82

In Table 7, AIV, SIV, and CIV share considerably similar amino acid sequence. The first table represents representative rules of each class and their frequency. Second table notice the percentage which AIV,SIV,CIV refers class1,2,3 with tat_window.

Table 8. Rule extraction under 7 window of TAT.

class	rule	frequency
class1	pos6=I pos7=L	0.80
class2	pos6=L pos7=I	0.75
class3	pos2=W pos6=V pos7=K	0.75

In Table 8, AIV is similar with SIV. SIV has similar amino acid with AIV. CIV is relatively different from others.

DOI: 10.1051/matecconf/20166901007

Table 9. Rule extraction under 9 window of TAT.

class	rule	frequency
class1	pos2=L pos6=W pos4=D pos9=R pos5=E pos8=R	0.80
class2	-	0.75
class3	pos2=I pos4=T	0.75

In Table 9, AIV is similar with SIV and CIV. SIV is similar with AIV. CIV is Smilar with AIV.

Table 10. Rule extraction under 5 window of VIF.

class	rule	frequency
class1	pos2=L pos4=R	0.80
class2	pos2=T	0.75
class3	pos4=W	0.75

In Table 10, AIV, SIV, and CIV share considerably similar amino acid sequence.

Table 11. Rule extraction under 7 window of VIF.

Class	rule	frequency
class1	pos2=L pos6=W pos4=D pos9=R pos5=E pos8=R	0.80
class2	-	0.75
class3	pos2=I pos4=T	0.75

In Table 11, AIV is different from others. However, SIV, and CIV are also different but not clearly.

Table 12. Rule extraction under 9 window of VIF.

class	rule	frequency
class1	pos1=F	0.75
class2	pos1=E pos9=G	0.75
class3	pos9=N	0.80

In Table 12, AIV is similar with SIV. SIV and CIV are similar each other as well as similar with AIV.

4.2 SVM

The experiments were held for each protein of 5window, 7window and 9window. In each result, we calculated the mean value of the data and drew the bar graph to enhance the visual recognition with the relative value. [16]

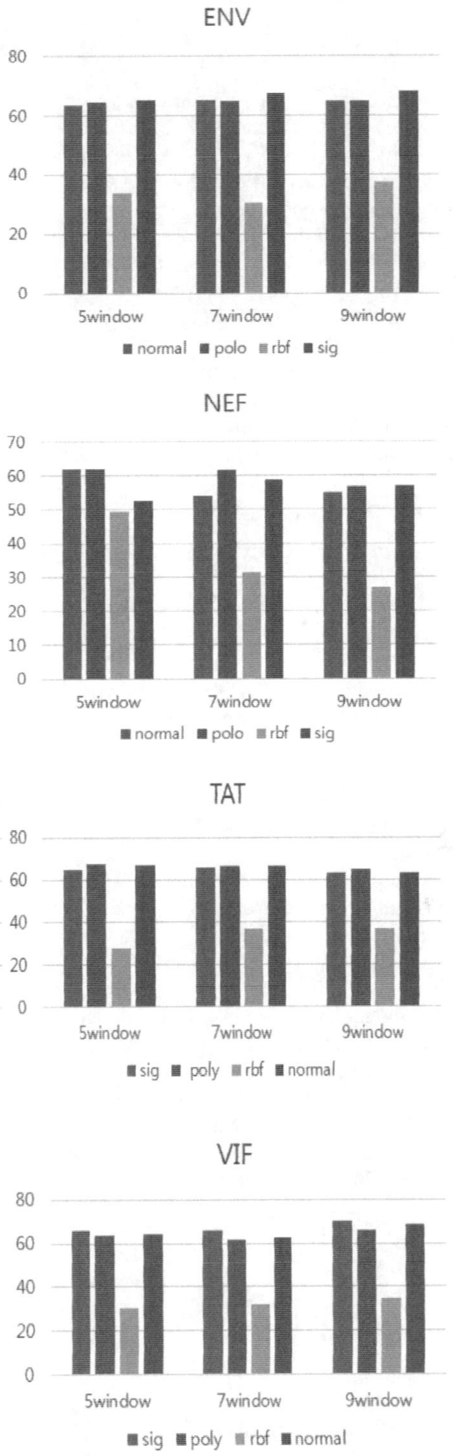

Figure 1. Mean value under 5 window, 7 window, 9 window of SVM

In Fig. 1, the average value of the rbf is evidently lower than that of normal, polynoml and signoid. The results means that the relation between the three viruses. SIV, CIV and AIV are in nonlinear relationship, so the three are seemed to be quite mixed.

5 Conclusion

The paper applied the approach of comparing the properties of CIV, SIV and AIV. In order to find out the genetic information of AIV, the 4 kinds of protein such as nef, tat, vif and env protein were the subjects of the experiments. Decision tree and SVM were used in the experiments. According to the rule which were extracted form the decision tree, the three viruses are evenly mixed in considerable amount of DNA sequences. From SVM, we realized that the relation between SIV, AIV and CIV are below the unlinear relationship. This similarities and differences discovered through our experiment may support other studies of the treatment of not only CIV but also SIV and AIV. We hope this study to be extended to other fields also in logical way and content aspects.

References

1. Colin R. Parrish, Yoshihiro Kawaoka (2005)."The Origins of New Pandemic Viruses: The Acquisition of New Host Ranges by Canine Parvovirus and Influenza A Viruses", Annual Review of Microbiology.
2. Characterization of a Novel Influenza A Virus Hemagglutinin Subtype (H16) Obtained from Black-Headed Gulls.
3. Clinical Signs and Symptoms Predicting Influenza Infection; Arnold S. Monto, MD; Stefan Gravenstein, MD; Michael Elliott, MD; Michael Colopy, PhD; Jo Schweinle, MD.
4. Chulseung Lee, Kwonil Jung, Jinsik Oh, Taehoon Oh, Sangyoon Han, Jeongmin Hwang, Minjoo Yeom, Donghyun Son, Jongman Kim, Bongkyun Park, Hyoungjoon Moon, Daesub Song ,Bokyu Kang(2009)."Protective efficacy and immunogenicity of an inactivated avian-origin H3N2 canine influenza vaccine in dogs challenged with the virulent virus", Vet Microbiol.
5. Shuo Su, Ye Chen, Fu-Rong Zhao, ji-Dang Chen, Jie-Xiang Xie, Zhong-Ming Chen, Zhen Huang, Yi-Ming Hu, Min-Ze Zhang, Li-Kai Tan, Gui-Hong Zhang, Shou-Jun Li(2013). "Avian-origin H3N2 canine influenza virus circulating in farmed dogs in Guangdong, China", Infection, Genetics and Evolution' Volume 19, Pages 251–256, October 13
6. Human influenza A H5N1 virus related to a highly pathogenic avian influenza virus
7. Wentworth DE1, McGregor MW, Macklin MD, Neumann V, Hinshaw VS. "Transmission of Swine Influenza Virus to Humans after Exposure to Experimentally Infected Pigs" ; Departments ofPathobiological Sciences and Veterinary Sciences,UniversityofWisconsin-Madison, and Agracetus, Inc., Middleton, Wisconsin
8. Webby RJ1, Swenson SL, Krauss SL, Gerrish PJ, Goyal SM, Webster RG. "Evolution of Swine H3N2 Influenza Viruses in the United States"
9. White JM, Hoffman LR, Arevalo JH, et al. (1997). "Attachment and entry of influenza virus into host cells. Pivotal roles of hemagglutinin". In Chiu W, Burnett RM, Garcea RL. Structural Biology of Viruses. Oxford University Press. pp. 80–104.
10. Heo, C. and Yoon, T. (2014) Deeper Understanding about Attributes of HIV Employing Support Vector Machine. International Journal of Bioscience, Biochemistry and Bioinformatics, 4, 336-339.
11. Lim, S.J., Heo, C., Hwang, Y. and Yoon, T. (2015) Analyzing Patterns of Various Avian In-fluenza Virus by Decision Tree. International Journal of Computer Theory and Engineering,7.
12. Jang, S. P., Park, K. H., Kim, Y. L., Cho, H. N., & Yoon, T. S. (2015). Comparison of H5N1, H5N8, and H3N2 Using Decision Tree and Apriori Algorithm. Journal of Biosciences and Medicines, 3(06), 49.
13. Go, E., Lee, S. and Yoon, T. (2014) Analysis of Ebolavirus with Decision Tree and Apriori Algorithm. International Journal of Machine Learning and Computing, 4.
14. Safavian, S. Rasoul, and David Landgrebe. "A survey of decision tree classifier methodology." IEEE transactions on systems, man, and cybernetics 21.3 (1991): 660-674.
15. Friedl, Mark A., and Carla E. Brodley. "Decision tree classification of land cover from remotely sensed data." Remote sensing of environment 61.3 (1997): 399-409
16. Furey, Terrence S., et al. "Support vector machine classification and validation of cancer tissue samples using microarray expression data." Bioinformatics 16.10 (2000): 906-914.

DOI: 10.1051/matecconf/20166901008

Deeper understanding about the genetic structure of dengue virus using SVM

Subin Choi, Hyunjin Choi and Taeseon Yoon

Natural Sciences, Hankuk Academy of Foreign Studies, HAFS, Yongin-si, Republic of Korea.

Abstract. Dengue fever, mainly found in the tropical and subtropical regions, is carried by mosquitoes. With the help of greenhouse effect, places considered to be a Dengue safe-zone are becoming more and more dangerous. Dengue fever shows similar aspects to MERS, which caused heavy casualties in South Korea; Dengue virus does not have clear treatments nor vaccines like MERS. Development of Dengue vaccine is actively investigated lately. However, it is not easy to succeed; the fact that Dengue's 4 serotypes have different properties and that repeated infections worsen the symptoms. This research aims to analyze the 4 serotypes (DENV1, DENV2, DENV3, DENV4) using SVM and ANN algorithms to investigate the constraints in the development of Dengue's vaccines and treatments.

1 Introduction

Dengue fever is an acute idiopathic disease caused by the a dengue virus which belongs to the Flavivirus family. Denge virus is native to the Southeast Asia, and is spread all over the tropical and subtropical regions around the world. Dengue becomes prevalent periodically and causes many fatalities. With the help of greenhouse effect, mosquitoes' population has proliferated (dengue is carried by mostquitoes) and it is continuously developing its influence.

In 2008, Ae. albopictus which is one of the mosquitoes that communicates Dengue virus was found in Jeju Island, South Korea. Also, 8 out of 35 Korean volunteers were infected at Colombo, Sri Lanka. South Korea is currently at risk since it was considered as a safe zone from Dengue fever. The fact that Dengue doesn't have clear treatments and vaccines classifies Dengue as a high-risk disease. Dengue virus has 4 serotypes (DENV1, DENV2, DENV3, DENV4). Even if DENV1 infections are fully cured, reinfection of a DENV2, DENV3, DENV 4 can be fatal [1]. This paper compares and analyzes the 4 serotypes using SVM and Neural Network Algorithm to investigate what delayed the development of the vaccines and to suggest development direction of the vaccine based on the similarities and differences of the 4 serotypes.

Dengue virus has 4 serotypes; DENV1, DENV2, DENV3, DENV4. Like the Dengue virus, Herpesvirus has about 80 different types of serotype. Out of these, eight of them are expressed on humans. Unlike the Dengue virus, where different vaccines are used to prevent the illness for the 4 distinct serotypes, Herpesvirus' sertoypes show different symptoms and are expressed in different parts of the body.

2 Materials and procedures

2.1 Dengue virus

Since Dengue Virus has been found in the 1950s in Philippines and Thailand, cases are constantly reported in most of Asia and Latin America. It is usually carried by a mosquitoes; Aedes aegypti, Ae. albopictus, Ae. polynesiensis, Ae. pseudoscutellares [2]. Dengue virus belongs to Flaviviridae, Flavivirus family and has 4 types of different serotyes; DENV1, DENV2, DENV3, DENV4 [3].

The structure of the gene is similar to those of the Flavivirus family. It has about 100 bases at both 5'-end and 3'-end, and noncoding regions (NCR) that contain about 600 bases. Proteins are expressed by a single open reading frame (ORF). 3 structural proteins (capsid (C), premembrane (prM), envelope (E)) and 7 nonstructural proteins (NS1, N2a, NS2b, NS3, NS4a, NS4b, NS5) are manufactured since the start of the translation stage and during the post-translation processes.

Dengue Virus 1 (DENV1) is most common among the 4 serotypes. Infections by DENV1 have the most serious symptoms[4],[5],[6]. Recently, Dengue virus infections are caused by superinfections or by the genetically mutated viruses rather than monoinfections.

2.2 SVM

SVM(Support Vector Machine), a field of Machine learning, is a model of supervised learning used for pattern recognition and data analysis. It is an approximated implementation which minimizes the structural jeopardy of the previous algorithms such as neural network algorithm [7], [8]. Data of a Test error (its range is decided by the sum of the learning errors' ratio) and a value which relies on the VC-dimension of a learning machine. It is generally used for classification and regression analysis, and is composed of hyperplane or its sets.

SVM divides the data into two groups to analyze with a Maximum margin hyper plane. Each vectors or points created at this stage is connected to form a polygon; convex hull [9]. These lines or polygons serve as a tool to divide the data into multiple groups and to compare the accuracy. The higher the accuracy, the two groups have considerable similarities. The lower the accuracy, the two have quite different properties.

2.3 Procedure

Genetic sequences of DENV1, DENV2, DENV3, DENV4 are extracted from the National Center for Biotechnology Information(NCBI). Experiments are carried out 10 times (10 fold cross validation) using SVM's main 4 functions; normal, poly, RBF, and sig. Each sequences of DENV1, DENV2, DENV3, DENV4 is divided into 10 groups and one from each is selected. The unselected 9 groups are put into the SVM algorithm and decide an invariable number representing the 'difference'. With this number, a single pre-selected group is classified using 4 functions, and the accuracy of the classfication vector is calculated. This procedure is repeated for 10 groups. In other words, 10 experiments are done per function. 9 window, 13window, 17 window is decided on how many sections the data are divided. The experiments are repeated 3 times per window. For each window, the average of the 4 functions' (sig, RBF, normal. poly) accuracy is found. These values are used to figure out the properties of the classification.

3 Results

For each 4 functions (normal, poly, RBF, sig), the experiments were repeated for each window (9 window, 13 window, 17 window). Normal function is involved with the linearlity; poly uses polynomial to figure out the nonlinear part; Gaussian function is used for RBF to classify the nonlinears; sig uses sigmoid function to classify the linear sections. DENV1, DENV2, DENV3, DENV4 can be classified using the function that has the highest average of accuracy. The figures below are the average of accuracy for each window.

3.1 9 window

We can recognize the similarities between the values of sig, normal and poly. The values of sig, normal and poly are 76.03498, 74.24741, 74.46065. When the sequences were classified with the 3 functions, the accuracy of the calssifcation vectors were relatively high; all functions' data were higher than 74%. Using these data, we could figure out that each serotyypes have significant differences in their properties (see Figure 1).

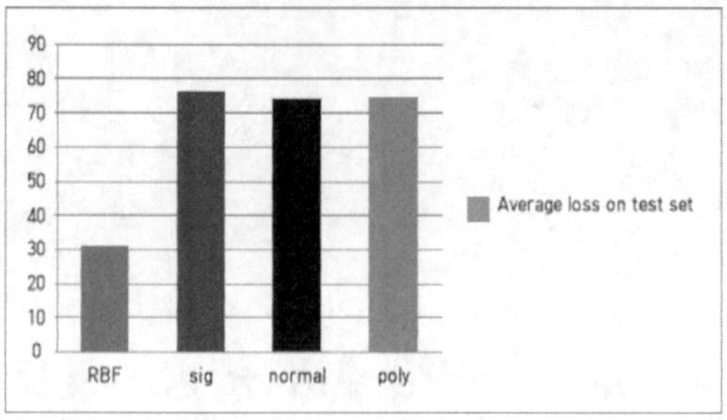

Figure 1. Average of accuracy for 9 window.

3.2 13 window

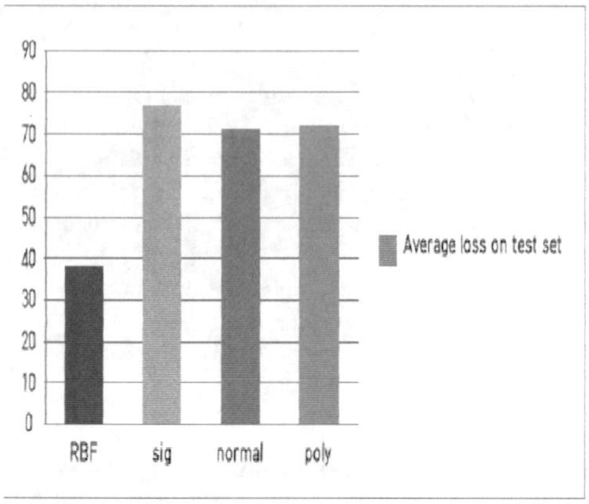

Figure 2. Average of accuracy for 13 window.

According to the results above, it is easy to recognize that sig, normal, poly has higher results than RBF. Therefore, it is obvious that when the sequences of Dengue fever were classified with 13 window, it can be divided nonlinearly and linearly (see Figure 2).

3.3 17 window

The result of the 17 window shows similar data with the 2 experiments above. Interestingly, the data of normal form the 3 experiments show values over 70%. Through the last experiment, we could confide that the classification of the Dengue virus is possible using the 4 functions (see Figure 3).

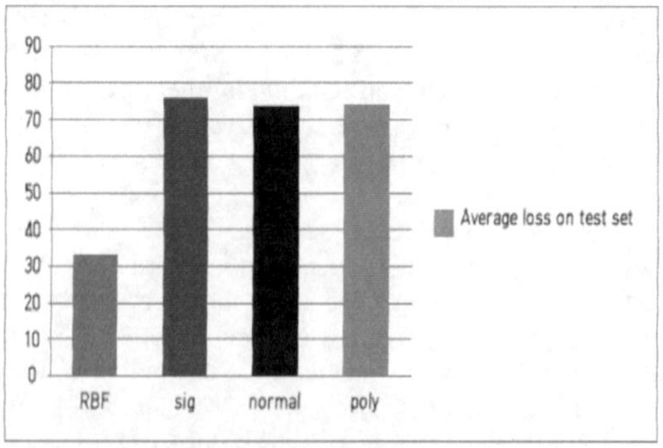

Figure 3. Average of accuracy for 17 window.

4 Analysis and discussion

Despite the risks of Dengue virus infections, there are few attempts to develop either the vaccine or the remedy (treatment) for Dengue virus. The most suitable treatment for the Dengue should satisfy the following qualifications. 1) The antibodies for the 4 different serotypes of Dengue should last for a long term. 2) It should stimulate the humoral immunity system and the cell-mediated immunity system fully.

We classified Dengue's 4 serotypes (DENV1, DENV2, DENV3, DENV4) with SVM algorithm using RBF, sig, normal and poly functions. Each results are organized by the number of the windows and the average values were compared. We could notice that the accuracy of sigmoid and polynomial functions were over 70%, and the accuracy of RBF function was relatively low; less than 40%. Therefore, it was obvious that DENV serotypes can be categorized both linearly and nonlinearly. For the normal function, DENV viruses showed accuracy of more than 70%. (However, it is comparatively lower than other viruses', which have the accuracy of over 90%).

As Dengue viruses are divided into 4 different groups, the Dengue vaccine should have the capability of preventing all types of the Dengue [10]. In addition, the antibodies for each serotype should be sustained homogenously and should irritate both humoral immunity system and the cell-mediated immunity system. For these reasons, vaccine for Dengue is tough to develop.

If the vaccine cannot properly prevent the Dengue for the 4 serotypes, there are 2 things we can think about. 1) The made vaccine doesn't work on certain serotypes (out of the 4 serotypes) 2) The vaccine has side effects. It can cause an additional infection when there are antibodies for other serotypes. It can make the situation worse; it causes Dengue fever and Dengue shock.

Accordingly, Dengue virus can be assorted by their serotypes and have different properties, but they also have few common features, to a certain extent. Although DENV1, DENV2, DEVN3, DENV4 show similar characteristics, according to the accuracy of each functions, we can conclude that different vaccines are needed for the different serotypes.

It is possible for a Dengue vaccine to induce more serious diseases. Therefore, the vaccine should be carefully developed, and should satisfy all qualifications of the 4 different serotypes of the Dengue fever, which can prevent counterproductive effects of using the vaccine to the patients.

5 Conclusion

Dengue virus, which causes the Dengue fever, has four different serotypes; DENV1, DENV2, DENV3, DENV4. According to the classification of each serotypes using SVM, the different serotypes do have similar features. Nevertheless, the sigmoid and polynomial functions' vector accuracy values are over 70% which means that the serotypes have distinct characteristics. Thus, from these results, we could draw the fact that different vaccines have to be used according to the different serotypes of the Dengue virus, even though they express similar symptoms of Dengue fever. Future vaccine developments have to be focused based on the distinct properties of each serotypes, not on the Dengue virus itself.

However, according to the results of this paper, it turned out that even though the Dengue serotypes show 70% differences, they also have similar properties. The 4 serotypes of Dengue virus (DENV1, DENV2, DENV3, DENV4) have different characteristics and can be classified into distinct groups. With the results of this paper, future research on the common reactions of each antibody on the different serotypes when Dengue is expressed and on the methods of developing a vaccine that efficiently uses the common properties of the 4 serotypes of Dengue virus will present the development course of the vaccines for the Dengue virus.

References

1. Sessions, October M., Andreas Wilm, Uma Sangumathi Kamaraj, Milly M. Choy, Angelia Chow, Yuwen Chong, Xin Mei Ong, Niranjan Nagarajan, Alex R. Cook, and Eng Eong Ooi. "Analysis of Dengue Virus Genetic Diversity during Human and Mosquito Infection Reveals Genetic Constraints." *PLoS Negl Trop Dis PLOS Neglected Tropical Diseases* 9.9 (2015): n. pag. Web.
2. Mondini, A., A. Terzian, A. Ferraz, F. Oliveira, F. Chiaravalloti Neto, and M. Nogueira. "Surveillance of DENV in a City from São Paulo from 2006 to 2011: The Emergence of DENV-3 and DENV-4 and the Reemergence of DENV-2 and DENV-1." International Journal of Infectious Diseases 16 (2012): n. pag. Web.
3. Kim, Hyunseong, Juyoung Yoo, and Taeseon Yoon. "An Analysis of the Genomes of Dengue Virus Using Decision Tree and Apriori Algorithm." International Journal of Computer and Communication Engineering IJCCE **5.4** (2015): 294-301. Web.
4. Thu, Hlaing Myat, Kym Lowry, Thein Thein Myint, Than Nu Shwe, Aye Maung Han, Kyu Kyu Khin, Kyaw Zin Thant, Soe Thein, and John Aaskov. "Myanmar Dengue Outbreak Associated with Displacement of Serotypes 2, 3, and 4 by Dengue 1." Emerg. Infect. Dis. Emerging Infectious Diseases 10.4 (2004): 593-97. Web.
5. Melo, Fernando Lucas De, Camila Malta Romano, and Paolo Marinho De Andrade Zanotto. "Introduction of Dengue Virus 4 (DENV-4) Genotype I into Brazil from Asia?" PLoS Negl Trop Dis PLoS Neglected Tropical Diseases 3.4 (2009): n. pag. Web.
6. Vaughn, David W., Sharone Green, Siripen Kalayanarooj, Bruce L. Innis, Suchitra Nimmannitya, Saroj Suntayakorn, Timothy P. Endy, Boonyos Raengsakulrach, Alan L. Rothman, Francis A. Ennis, and Ananda Nisalak. "Dengue Viremia Titer, Antibody Response Pattern, and Virus Serotype

Correlate with Disease Severity." The Journal of Infectious Diseases J INFECT DIS 181.1 (2000): 2-9. Web.
7. "Support Vector Machine (SVM)." SpringerReference (n.d.): n. pag. Web.
8. Law, Ka-Man. "Vaccine Development against the Severe Acute Respiratory Syndrome-coronavirus (SARS-CoV) Using SARS-CoV Spike Protein." (n.d.): n. pag. Web.
9. Zeng, Jianwu, and Wei Qiao. "Short-term Solar Power Prediction Using a Support Vector Machine." Renewable Energy 52 (2013): 118-27. Web.
10. Gubler, Duane. "Dengue/Dengue Hemorrhagic Fever: The Emergence of a Global Health Problem." Emerg. Infect. Dis. Emerging Infectious Diseases 1.2 (1995):55-57.Web.

 DOI: 10.1051/matecconf/20166902001

Electricity generation from the mud by using microbial fuel cell

Sitinoor Adeib Idris, Farah Nasyitah Esat, Ain Adilla Abd Rahim, W. A. Zahin Rizzqi W. Ruzlee and W. M. Zyaid Razali

Faculty of Chemical Engineering, Universiti Teknologi MARA, Johor, Pasir Gudang Campus

Abstract. Microbial fuel cells (MFCs) is a bio-electrochemical device that harnesses the power of respiring microbes to convert organic substrates directly into electrical energy. This is achieved when bacteria transfer electrons to an electrode rather than directly to an electron acceptor. Their technical feasibility has recently been proven and there is great enthusiasm in the scientific community that MFCs could provide a source of "green electricity". Microbial fuel cells work by allowing bacteria to do what they do best, oxidize and reduce organic molecules. Bacterial respiration is basically one big redox reaction in which electrons are being moved around. The objective is to generate electricity throughout the biochemical process using chemical waste basically sludge, via microbial fuel cells. The methodology includes collecting sludge from different locations, set up microbial fuel cells with the aid of salt bridge and observing the results in voltage measurement. The microbial fuel cells consist of two chambers, iron electrodes, copper wire, air pump (to increase the efficiency of electron transfer), water, sludge and salt bridge. After several observations, it is seen that this MFC can achieve up until 202 milivolts (0.202volts) with the presence of air pump. It is proven through the experiments that sludge from different locations gives different results in term of the voltage measurement. This is basically because in different locations of sludge contain different type and amount of nutrients to provide the growth of bacteria. Apart from that, salt bridge also play an important role in order to transport the proton from cathode to anode. A longer salt bridge will give a higher voltage compared to a short salt bridge. On the other hand, the limitations that this experiment facing is the voltage that being produced did not last long as the bacteria activity slows down gradually and the voltage produced are not really great in amount. Lastly to conclude, microbial fuel cell essentially is a solution for a renewable energy emitted by bacteria activity that need to be take a further attention , research and development

1 Introduction

Microbial fuel cells (MFCs) use bacteria to convert the chemical energy of a particular substrate contained in wastewater into electrical energy. This is achieved when bacteria transfer electrons to an electrode rather than directly to an electron acceptor. Their technical feasibility has recently been proven and there is great enthusiasm in the scientific community that MFCs could provide a source of "green electricity" by exploiting domestic and industrial waste to generate power [1]-[3]. With the usage of mud which clearly contain bacteria in it, it is the main factor of the selection of mud for this microbial fuel cell. Bacteria need energy to survive, in the same way that humans need food to live. Bacteria get this energy in a two-step process. The first step requires the removal of electrons from some source of organic matter (oxidation), and the second step consists of giving those electrons to something that will accept them (reduction), such as oxygen or nitrate. If certain bacteria are grown under anaerobic conditions (without the presence of oxygen), they can transfer electrons to a carbon electrode (anode). The electrons then move across a wire under a load (resistor) to the cathode where they combine with protons and oxygen to form water. When these electrons flow from the anode to the cathode, they generate the current and voltage to make electricity [4]. Microbial fuel cells work by allowing bacteria to do what they do best, oxidize and reduce organic molecules. Bacterial respiration is basically one big redox reaction in which electrons are being moved around. Whenever a moving electrons present, the potential exists for harnessing an electromotive force to perform useful work [1].

As to assist the flow of electricity, the presence of salt bridge is crucial as it transports the proton in microbial fuel cell (MFC). In order to offset the very high cost of proton exchange membrane, salt bridges have been used in dual chamber MFCs [5]. The parameters that has been selected for the experimental procedure in this microbial fuel cell (MFC) are upon two factors which are; type of sludge and the length of salt bridge. It is expected that different type of sludge gives different result in term of the voltage in electricity. This is because from different location, it gives varies type of bacteria in the mud. The length of salt bridge also effects the generation of electricity because of the efficiency of transportation of proton [5].

The most immediately foreseeable application of an MFC is in waste water treatment. Microbes love sewage, and the conditions of a waste water treatment plant are ideal for the types of bacteria that can be used in an MFC. Exoelectrogens are more than happy to breakdown and metabolize the carbon rich sewage of a waste water stream to produce electrons that can stream into a cheap conductive carbon cloth anode. The electricity generated from the MFC also offsets the energy cost of operating the plant. As an added bonus, the bacteria eat a lot of the sludge normally present in waste water. One of the company that apply this process claims to be able to cut sludge down by 80% in their waste water treatment processes, which saves them time and money from having to transport sludge to a landfill or wasteland [1].

The application of MFC in integrated strategies towards sustainable bioenergy generation is current interest. MFCs are specifically designed to harness bioelectricity from benthic aquatic eco-system using natural habitats wherein, ecological water bodies are considered as a potential source to convert the organic rich sediment to energy. Plant based MFCs helps in the utilization of solar radiation to generate bioelectricity by integrating the rhizo deposits of living plant with electrodes. The scope of MFC application has been extended to the development of photo-electrocatalytic fuel cell using photosynthetic organisms as anodic biocatalyst to assess their potential to harness power [6].

The limitations of microbial fuel cell in this experiment is it only consists of two chambers and the amount of mud in each chamber which approximately 1.5L. The result that is predicted to be obtained is up to 200mV. The performance is severely affected based on irreversible reactions and processes in the anode and the cathode compartments.

2 Methodology

2.1 Materials

Thick, soft, wet mud or a similar viscous mixture of liquid and solid components which contain polyphosphate bacteria is collected. The bacteria that present in this sludge carries out break down of nutrients and organic matter inside the sludge leads toward emission of electron which is particularly a contribution of small volts current. Two types of sludge from different location are collected as to determine the maximum performance of the sludge for the microbial fuel cell. First sample was collected at the bottom of the lakeside while the second sample was collected at the creek also from lakeside. The mud that are collected had different physical traits such as the colour and the roughness. Two containers were needed as the chamber of the microbial fuel cell. Both lids of containers were drilled for the insertion of copper wire that been attached with iron mesh electrode. Next, the wall of the containers were also drilled a hole each, for salt bridge connection. The salt bridge was prepared basically from these materials; cotton rope, salt and water. An amount of 98g of salt was firstly being dissolved in water before it is heated for approximately 2 minutes. The cotton rope is soaked into the salt water for a period until the salt water has fully absorbed by the cotton rope. Salt bridge is a medium for proton to be flown from water filled container while electron counter flow by the copper wire. The electrode that were used in this experiment is iron mesh gauze as it is a good electricity conductor. The iron mesh were folded a few times and they were bind together to the both end of the copper wire that connect the two chambers. The iron electrode functions to attract electron release by the bacteria and flow it to copper wire.

2.2 Procedure

First, one of the container was filled with water and another container was filled with sludge. Then, the iron electrode were submerged into both of the container which contained water and sludge. Next, the lids of the container were closed and the aerobic bacteria container should be exposed to small amount of oxygen. After that, the copper wire at the both container were attached to alligator clips, then the alligator clips were clamped to the multimeter probes. The black probes should be connected to the electrode at the sludge where the electron emitted from the sludge were attracted by the anode. The current flow from the sludge container were detected and measured by the multimeter in volt measurement. Then, an air pump was turn on to increase the efficiency of electron transfer. Fig. 1 shows the diagram of how the microbial fuel cell was set up.

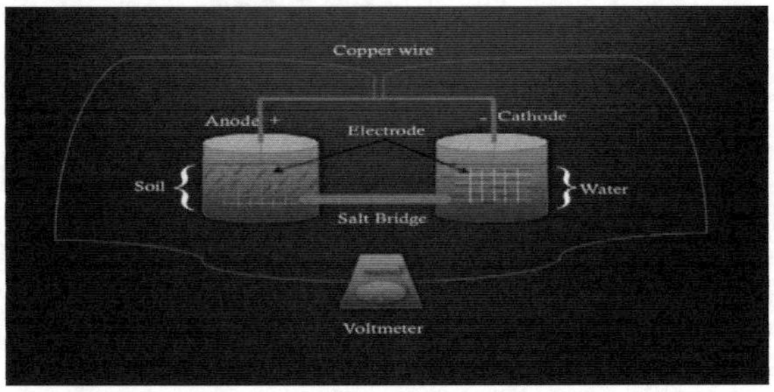

Figure 1. The set up of microbial fuel cell

3 Results and discussion

There were two parameters recorded as to determine the effect of each parameter towards the electrical charges produces from microbial fuel cell which are, location or type of sludge and length of salt bridge. Apart from that, the presence of air pump in the microbial fuel cell also give a huge impacts to the results obtain. Air pump was used as to increase the efficiency of the electrical charges produce from bacteria as it will supply more oxygen to the biotic ecosystem. Bacteria also known as aerobic which need oxygen to yield energy for growth. Different location of sludge will provide different amount of electrical charges form as in each sludge it contains different amount and type of nutrient. Fig. 2 and Fig. 3 shows the results which proves the difference in each sludge and length of salt bridge with the presence of air pump give different electrical measure in millivolt (mV).

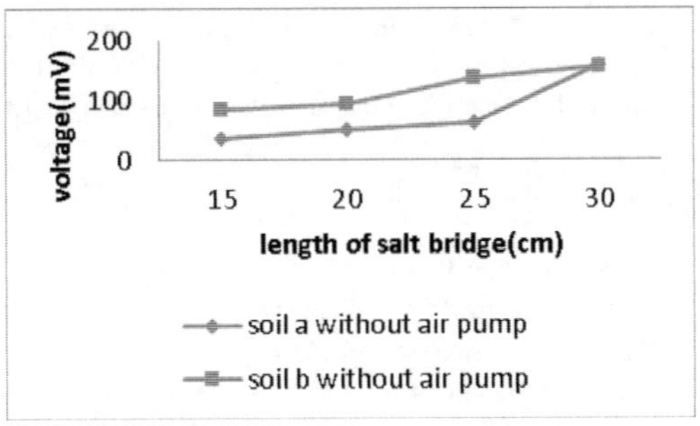

Figure 2. Electrical charges (mV) to the length of salt bridge (cm)

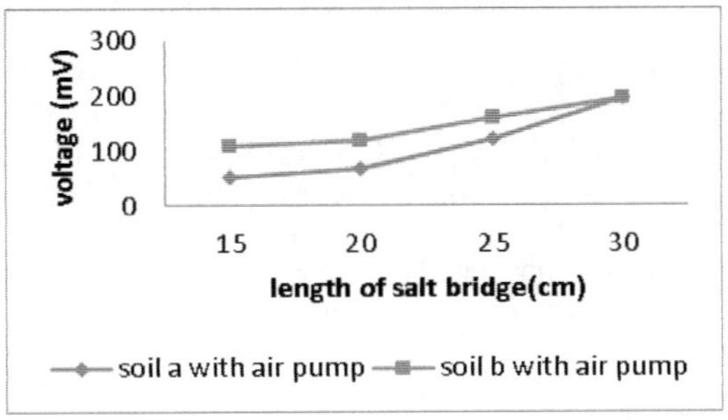

Figure 3. Electrical charges (mV) to the length of salt bridge (cm)

Fig. 2 shows that sludge B produced higher amount of electrical charges rather than sludge A in every length of salt bridge. This results are proven that it follows the theory as the maximum amount of electrical charges can be convert is 789mV as it's depends on the substrate use in the experiment [7], [8]. The presence of added oxygen air pump also gives out a lot of effects towards the microorganism performances. Based on the Fig. 2 and 3 it shows a lot of differences in the electrical charges detected by the multimeter. It was proved that bacteria need more oxygen as energy to growth for their development. Thus from the reaction between the microorganism, an electrical charges will be form [1]. Sludge A was collected in different location as well as sludge B. This different location parameter will gives out different amount of microorganism or bacteria contain in each sludge.

Each sludge in different location will have different type of microorganism contain in it thus every type of microorganism will give different performance to the microbial fuel cell. In a review it stated that, axenic culture (single bacterial species) such as *Proteus vulgaris* will give out 0.8mA while *Rhodoferax ferrireducens* from the same species will give out only 0.2mA for the performances [7]. Microbial fuel cell were operated by using the concept of microorganism as the catalyst. This concept were shown practically in the waste water treatment plant. In waste water treatment plant, a specific type of microorganism will be supply to the plant with oxygen to improve the effectiveness of the process. As different microorganism will provide different performances, the pathway or the type of electrode connected to the substrate also will affect the microorganism performance [8].

The absence of any acceptors in an MFC makes some microorganisms pass the electrons onto the anode surface. Difference in redox potentials (i.e., the ability of a compound to donate or accept electrons, denoted E_o and measured in volts) between the electron donor and the electron acceptor is the determinant of the potential energy available to the microorganism for anabolic processes. In an MFC the electrochemical redox potential difference of the anode and cathode determines how much energy is available and thus, generate the electricity [5]. By the results from the experiment, it is proven how redox reaction actually works with the presence of anode and cathode, by the assist of proton-transport which is the salt bridge.

The salt-bridge acts a proton transferring channel and this study aims to optimise the concentration of sludge that facilitates the maximum movement of the protons towards the cathode that corresponds with higher current generation efficiency of the MFC [3]. Therefore, the presence of salt bridge is crucial in order to ensure the transfer of protons work efficiently and eventually gives high electricity generation. After several experiments, it shows that different length of salt bridge give different results in terms of voltage. As well as shown in the Fig. 2 and 3, although different types of sludge were used during the experiment, it still give the highest electrical charges for both sludge when the highest length of salt bridge were use which is 30cm long.

The concentration of the salt bridge were recorded as constants as it shows that length of salt bridge can give huge effect toward the microbial fuel cell performances. Salt bridge were used to transfer the electron from cathode to anode as the longer the salt bridge, the higher the amount of electron will be transferred as it is proven during the experiments. The salt bridge with the length of 30cm in the presence of air pump gives out 190.5mV for sludge A and 192.7mV for sludge B while the microbial fuel cell with 15cm length of salt bridges gives out only 53.6mV for sludge A and 108.7 for sludge B.

The uses of microbial fuel cell in the waste water treatment offer several advantages such as cells are able to recover the chemical energy present in wastewater and turn it into electrical power [1]. This shows that the microorganism in the cell will react to convert chemical energy to the electrical energy. The sludge bacteria eat what is in the sludge, such as microscopic nutrients and sugars, and in turn, produce electrons that are released back into the sludge. Electrons are subatomic particles that have a negative charge which is used to create electricity, which is a form of energy. Most of the microorganism in microbial fuel cell depends on the oxogeneous or endogeneous mediator molecules transfer but a few can deliver electrons directly to the anode [4]. As in the experiment, which is using biofilm as the factor, microorganism can transfer electrons directly from the cell envelope to the electrode or across the biofilm.

The two chambers were stuck together tightly to prevent the salt bridge from getting loosed or deattached from the chamber. This is because the cotton rope that is used as salt bridge as it is easily deattached from the chamber when one of the chamber is moved and leakage happen. Thus, it will affect the result. Besides that, the used of the electrodes that have high surface area to increase the efficiency of the transfer of electron from the anode to the cathode. For example, in this experiment, the electrode that have 1 cm width is used. Therefore, by increasing the width to 2 cm is good enough to increase the voltage generated. The amount of power that is generated in these systems is affected by the surface area of the cathode relative to that of the anode and the surface of the membrane. For example, add more air pump to supply oxygen in cathode chamber. This can help in increasing the oxygen quantity in the water.

4 Conclusions and recommendations

The highest result for microbial fuel cell with the absence of air pump is 153mV. The difference shows that the higher amount of oxygen provided will produce a result that is more efficient. This microbial fuel cell's experiment that convert microorganisms activity inside mud into electricity by redox reaction was predetermined as one of a green solution as it have a possibility to instill a non-harmful energy and environmentally friendly. Systems beneath the microbial fuel cell would give out an extensive search of ways to operate the bio-electrochemical device in more efficient and in high energy production. This fuel cell are feasible with the water treatment process in harvesting energy utilized by anaerobic digestion as a method used in the microbial fuel cell to collect bioenergy from wastewater, from this scope, MFC had clearly give an advantages in treating wastewater linearly producing electricity.

The microbial fuel cell experiment that had been carried out had given a process on learning the flaws of the device and the data that been acquired help to improvise a small scale of microbial fuel cell to reach its optimal condition. Currently, the size of MFC's is limited by the fact that electron transport only occurs in a bacteria layer immediately in contact with the electrodes. So while MFC's would have been a success in large scale batch processing of waste water streams, their true potential actually lies in small scale devices where the surface to volume ratio is high. There exists an optimal flow rate of reactants for increasing the voltage output of an MFC. Advances in microfluidics will allow engineers to make increasingly smaller MFC devices that can take advantage of this high surface to volume ratio. Research into advanced microfluidics, bacterial strains, a more robust separator membranes and efficient electrodes are the key to unlocking the maximum potential of MFCs.

The power density produced by these systems is typically limited by high internal resistance and electrode-based losses. When comparing power produced by these systems, it makes the most sense to compare them on the basis of equally sized anodes, cathodes, and membranes. In term of internal resistance of the MFC, it is dependent on both the resistance of the electrolyte between the electrodes and by the membrane resistance. For optimal operation, anode and cathode need to be as close together as possible. Also proton migration significantly influences resistance-related losses [9-11].

References

1. F. J. Hernández-Fernández, Pérez de los Ríos, a., M. J. Salar-García, V. M. Ortiz-Martínez, L. J. Lozano-Blanco, C. Godínez, J. Quesada-Medina, Recent progress and perspectives in microbial fuel cells for bioenergy generation and wastewater treatment. Fuel Processing Technology, **138**, 284–297, (2015).
2. Y. Li, Y. Wu,B. Liu,H. Luan, T. Vadas, W. Guo. B. Li, Bioresource Technology Self-sustained reduction of multiple metals in a microbial fuel cell – microbial electrolysis cell hybrid system. BIORESOURCE TECHNOLOGY, **192**, 238–246. (2015).
3. W. Liu, S. Cheng, D. Sun, H. Huang, J. Chen, & K. Cen, Inhibition of microbial growth on air cathodes of single chamber microbial fuel cells by incorporating enrofloxacin into the catalyst layer. Biosensors & Bioelectronics, **72**, 44–50. (2015).
4. B. E. Logan, Microbial Fuel Cells. Retrieved from www.research.psu.edu/.../MFC_QandA.pdf, (2008).
5. T. R. Sreekrishnan and S. Sevda, Effect of salt concentration and mediators in salt bridge microbial fuel cell for electricity generation from synthetic wastewater. Retrieved from http://www.ncbi.nlm.nih.gov/pubmed/22423995, (2012).
6. S. V. Mohan, G. Velvizhi, J. A. Modestra & S. Srikanth, Microbial fuel cell : Critical factors regulating bio-catalyzed electrochemical process and recent advancements. Renewable and Sustainable Energy Reviews, **40**, 779–797. (2014).
7. K. Rabaey, & W. Verstraete, Microbial fuel cells: Novel biotechnology for energy generation. Trends in Biotechnology, **23(6)**, 291–298. (2005).
8. P. Aelterman, K. Rabaey, H. T. Pham, N. Boon and, W. Verstraete, Continuous electricity generation at high voltages and currents using stacked microbial fuel cells, Environmental Science and Technology, **40 (10)**, 3388–3394, (2006).
9. H. Wang,, H. Luo, P. H. Fallgren, S. Jin., & Z. J. Ren, Bioelectrochemical system platform for sustainable environmental remediation and energy generation. Biotechnology Advances, **33(3-4),** 317–334, (2015).
10. Y. Zhang,, Y. Wang., & I. Angelidaki, Alternate switching between microbial fuel cell and microbial electrolysis cell operation as a new method to control H_2O_2 level in Bioelectro-Fenton system. Journal of Power Sources, **291**, 108–116, (2015).
11. H. Rismani-Yazdi, Cathodic limitations in microbial fuel cells: An overview. Retrieved from http://www.sciencedirect.com/science/article/pii/S037877530800387X, (2008).

Biodiesel production from jatropha oil in a closed system

W. Shaaban[1,2], A. H. El-Shazly[1], M. F. Elkady[1] and M. Ohshima[3]

[1]*Chemical and Petrochemicals Engineering Department, Egypt-Japan University of Science and Technology, New Borg El-Arab City, Alexandria, Egypt*
[2]*Mechanical Power Engineering Department, Mansoura University, Egypt*
[3]*Chemical Engineering Department, Kyoto University, Kyoto, Japan*

Abstract. The use of biodiesel as an alternative fuel is becoming increasingly popular nowadays due to global energy shortage. The interest in using Jatropha as a non-edible oil feedstock is rapidly growing. Biodiesel produced from crude Jatropha oil with NaOH as a catalyst is investigated. Transesterification by methanol is carried out in a closed vessel as a batch system. Factors affecting the process which included the reaction temperature and pressure, reaction time, the molar ratio of methanol to oil and catalyst amount are investigated. The maximum conversion ratio of methyl ester yield of 97.7% was recorded under the conditions of 65 ºC, 1% (by mass) NaOH of the oil mass and 6:1 methanol to oil ratio.

1 Introduction

There is a growing enthusiasm for using renewable energies to reduce greenhouse effect as well as to reduce the dependence on petrol–based fuel because of the environmental, economic and supply troubles of fossil fuels [1]. Nowadays, when fossil fuel recourses are waning quickly, biodiesel fuels are turning out to be worthwhile. Biodiesel is a renewable and environmentally clean fuel for diesel engines [2]. Being environmentally beneficial, biodegradable, non-toxic and with low pollution emission makes biodiesel remarkable for utilization [3-5]. Distinctive ways have been accounted for biodiesel production such as micro-emulsification, pyrolysis, and transesterification [6]. The most widely recognized approach to delivering biodiesel is transesterification process, which refers to a catalyzed chemical reaction involving vegetable oil and alcohol to yield fatty acid alkyl esters (biodiesel) and glycerol as a by-product [6]. Methanol is the most usually utilized alcohol as a result of its low cost and its physicochemical advantages (polar compound and short chain alcohol) [7]. At the industrial scale, the alkaline catalyst is the most utilized because of its suitable operational conditions: low cost, simple to install and, above all, high reaction rate which gives a reduction in the equipment size [8].

Currently, edible oils are the main resources for world biodiesel production (more than 95%) [9]. However, the broad utilization of edible oils may prompt some undesirable impacts such as starvation and higher food prices in developing countries [10]. Hence, non-edible plant oils turn out to be exceptionally encouraging option feedstock for biodiesel production [11]. Non-edible vegetable oils which are known as the second generation feedstocks can be considered as a promising alternative for traditional edible food crops for the production of biodiesel. However, as a serious drawback, most non-edible oils contain a high content of free fatty acids (FFAs), which increases the biodiesel production cost [12-14]. There are an enormous number of oil plants that deliver non-edible oils. From a list of 75 plant species containing oil in their seeds or kernels more than 30%, 26 species were reported by Azam et al. [15] as potential sources of the oil that can be employed for the production of FAMEs suitable for the use as biodiesel. Jatropha curcas crude oil has been used for the biodiesel production [16]. Jatropha is a species of flowering plant in the spurge family, Euphorbiaceaea and it is a drought-resistant tree. It is a plant with numerous characteristics, different uses and extensive potential. The plant can be utilized to avoid and/or control disintegration, to recover area, developed as a live fence, especially to contain or exclude farm animals and be planted as a commercial crop. It is a native of tropical America, however now flourishes in numerous parts of the tropics and sub-tropics in Africa/Asia. It has few nuisances and diseases and will grow under a wide range of rainfall regimes from 200 to over 1500 mm per annum. In low rainfall areas and in delayed rainless periods, the plant sheds its leaves as a counter to dry spell [17].

Jatropha is easy to establish, grows relatively quickly and is hardy, being drought tolerant. It is not browsed, for its leaves and stems are poisonous to creatures, but after treatment, the seeds or seed cake could be used as a creature nourish. The oil from these seeds has low acidity, good oxidation stability, low viscosity and good cold properties [17]. Accordingly, this investigation was concentrated on the biodiesel production utilizing

Egyptian Jatropha curcas crude oil (JCCO) in a closed vessel as a batch system.

2 Experimental

2.1 Chemicals and materials

2.1.1 Chemicals

Methanol with purity of 99.9% (Alpha Chemical Company, Egypt), acetic acid (Alpha Chemical Company, Egypt), Ethanol (Wako Pure Chemical Industries, Japan) and sodium hydroxide (Sigma-Aldrich Company, USA) were used without further purification. Freshly distilled water was used during the washing process. Ethanol was used for preparation of all produced biodiesel samples for gas chromatography analysis.

2.1.2 Jatropha oil

Jatropha curcas is a poisonous, semi-evergreen shrub or small tree, reaching a height of 6 m [18]. It is resistant to a high degree of aridity, allowing it to be grown in deserts. The fruits are produced in winter, or there may be several crops during the year if soil moisture is good and temperatures are sufficiently high. Most fruit production is concentrated from midsummer to late fall with variations in production peaks where some plants have two or three harvests and some produce continuously through the season. The fruit is oval-shaped ranging in length between 1-2 cm (Fig. 1) and every one contain from one to three seeds. the seeds are mature when the capsule changes from green to black as shown in Fig. 2. The seeds contain 27-40% oil [19] (average: 34.4% [20]) that can be processed to produce a high-quality biodiesel fuel, usable in a standard diesel engine.

The crude material used in this work is acquired by getting the fruits of Jatropha from the trees growing in a governmental farm located in Luxor city (in upper Egypt) and squeezing the gathered seeds by a hydraulic piston. The residual particles in the oil are uprooted by filtration using 8μm filter paper under vacuum conditions. The separated oil is stored until the analysis of the oil composition and the biodiesel production procedures are performed (1 month).

Figure 1. A photo of the Jatropha fruits.

Figure 2. A photo of the Jatropha seeds.

2.2 Biodiesel production process

A schematic illustration of the experimental setup appears in Fig. 3 which utilized for transesterification reaction. The device is composed of a high-pressure stainless steel vessel (EV-3-50-2) of 50 ml capacity as a closed batch reactor with pressure display (as shown in Fig. 4). This vessel is made from stainless steel 316 and has a maximum temperature and pressure limits of 100 °C and 30 MPa , where the heating system consisted of a GC furnace. The amount of NaOH is dissolved in methanol then this was mixed with the oil in the closed vessel. The molar ratios of methanol to oil are adjusted from 4.5 to 9. After a specific time, the reaction termination is done by addition of the acetic acid solution (20%).

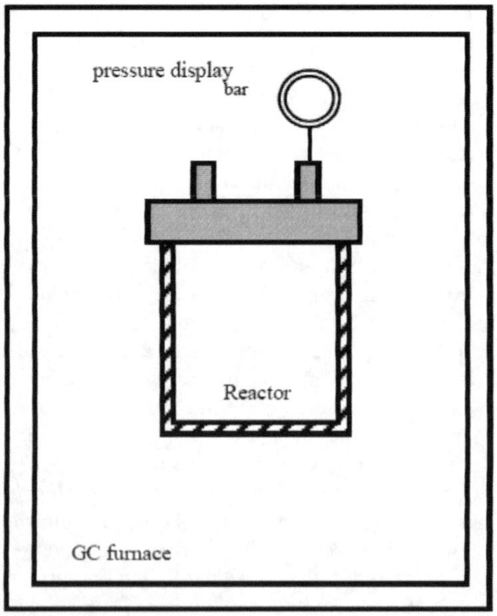

Figure 3. Experimental setup of the closed reactor.

The collected product is then centrifuged at 5000 rpm for 8 min. The upper fatty acid methyl ester (FAME) layer is cleaned several times with distilled water to dismiss the remaining inorganic components. 1 ml of the washed sample is mixed with 9 ml of methanol then filtered through a syringe filter (0.45 μm) to be analysed using a gas chromatography GC 2010 plus (AOC-20i).

Figure 4. A photo for the pressure vessel.

3 Result and discussion

3.1 Pressure-Temperature relation

Fig. 5 represents the relation between the temperature and the obtained manometric pressure at two values of filling ratio which represents the ratio between the volume of the total reactants and the reactor which in turn leads to an increase in pressure. It is clear that there is a direct correlation between the reaction temperature and pressure generated.

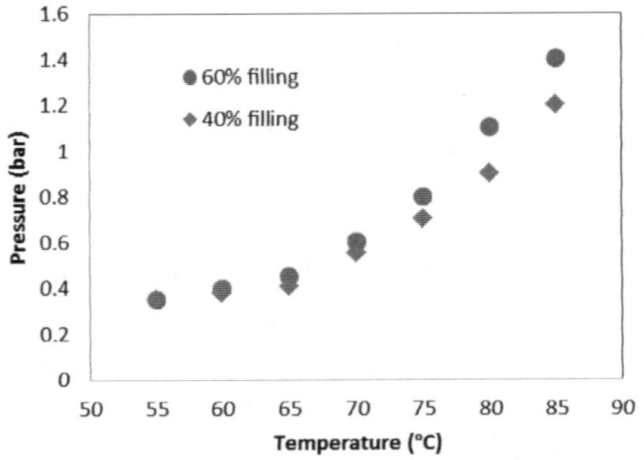

Figure 5. Relation between temperature and manometric pressure.

3.2 Effect of molar ratio

The theoretical methanol to oil ratio is 3:1 but higher molar proportion ought to be used to ensure that the reaction would be completed in light of the fact that transesterification reaction cannot be accomplished in that stoichiometric proportion. In this study, the molar ratio of methanol to oil is contrasted from 4.5 to 9. The impact of the molar ratio of methanol to oil on the percentage of methyl esters conversion is established in Fig. 6. In this figure, every point is recognized with the mean percentage of methyl esters for all keeps runs at the same molar ratio.

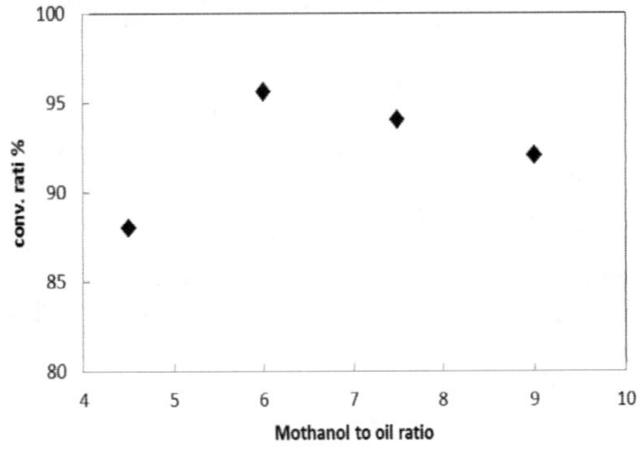

Figure 6. Effect of molar ratio on conversion ratio.

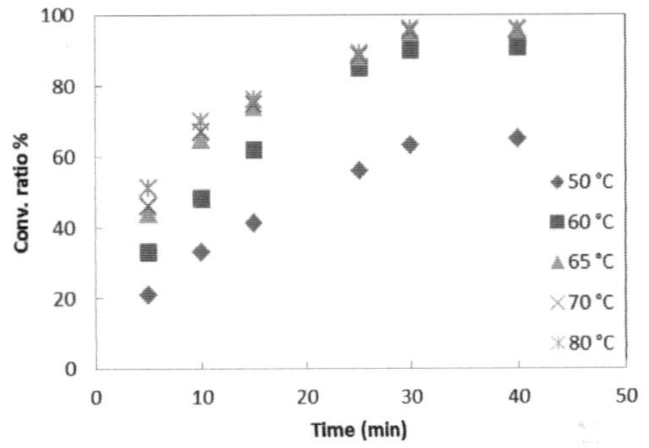

Figure 7. Effect of temperature on conversion ratio

The figure clarifies that the percentage of methyl esters expanded as the molar ratio of methanol expanded from 4.5:1 to 6:1. However, additional develop in the molar ratio of methanol from 6:1 to 9:1 brought about a slim lessening in the percentage of methyl esters. This behavior might be owed to the separation of glycerine and methyl ester. So, an extra molar ratio of methanol can diminish the percentage of produced methyl esters.

3.3 Effect of reaction temperature

The impact of the temperature on the percentage of methyl esters conversion is demonstrated in Fig. 7. The experiments are coordinated at reaction temperature range from 50 to 80 °C over residence time period from 5 to 40min. In this figure, every point affirms the Conversion ratio of methyl esters for all runs at the optimum conditions of molar ratio and the amount catalyst. It ought to be denoted that at low temperature, the reaction rate is slow and with expanding temperature, the percentage of methyl esters remarkably expanded.

The most extreme percentage of methyl esters for jatropha oil utilizing NaOH is done at 65 ºC. On the other hand, as far as the temperature of the reaction is higher than the boiling point of methanol in typical pressure condition, a further increment in temperature to 70 ºC caused the decline in the percentage of methyl esters. Furthermore, due to the high content of FFA in oil, increase in reaction temperature quickens the glycerides saponification by the alkaline catalyst.

3.4 Effect of the reaction residence time

The reaction time in biodiesel production is an exceptionally crucial parameter due to its role in the mass transfer between the two phases of reactants. Fig. 8 displays the effect of reactant residence time on the percentage of methyl esters conversion. The experiment is led at a temperature of 65 ℃, NaOH amount of 1 Wt.% of the oil and a molar ratio (methanol/oil) of 6. The outcomes assigned that percentage of methyl esters conversion expanded from 65% to 95.6 % when the residence time is expanded from 10 to 40min. The percentage of methyl esters conversion change initially expanded to a great degree with expansion in residence time from 10 to 30 s and after that expanded marginally. The most extreme estimation of the methyl ester achieved very nearly 30min and transesterification of oil is verging on complete. It can be observed that at residence time longer than 30min, the percentage of methyl esters conversion did not change eminently. Therefore, at the optimum condition with minimum residence time, a high percentage of methyl esters are accomplished with the most reduced biodiesel production cost.

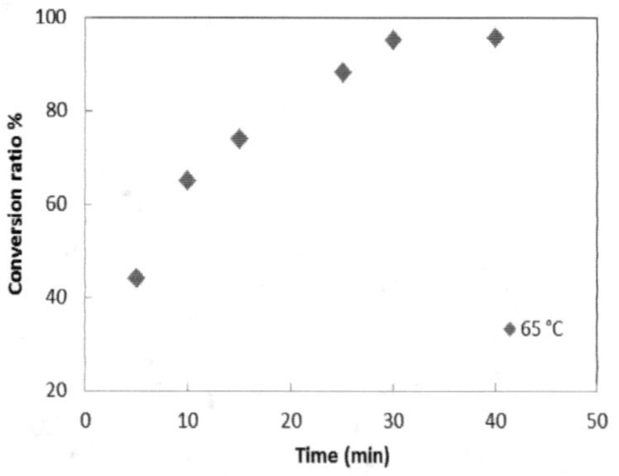

Figure 8. Effect of reaction residence time on conversion ratio

3.5 Comparison between the closed and the open system

Comparing the biodiesel production yield of Jatropha at closed and open system, it was found that the percentage of methyl esters conversion in the closed vessel is slightly better than that of the open vessel as shown in Fig. 9. May come to mind that the pressure plays a role in this increment, but as it known that increasing the pressure on a reaction involving only liquids has no effect on the rate. So, keeping methanol from escaping, in the case of the open system, may be the well explains this improvement. Increasing the amount of methanol in the reaction region is exactly the same as increasing the methanol to oil ratio.

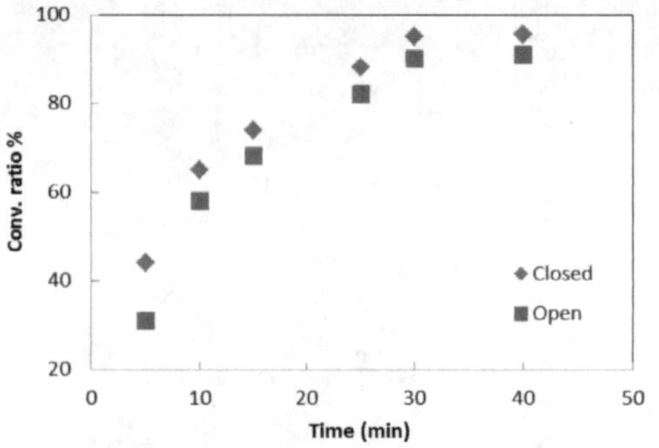

Figure 9. Comparison between the closed and the open system

4 Conclusion

The base catalyzed methanolysis of jatropha oil takes place at a closed vessel as a batch reactor. Factors affecting the process is studied and the highest conversion ratio of methyl esters conversion of 97.7% is obtained within 30min using 6:1 molar ratio of methanol to oil at the optimum conditions. It is shown that the yield of fatty acid ethyl esters depends strongly on mass transfer limitations. It is found that the maximum percentage of methyl esters for jatropha oil using NaOH is accomplished at 65 ℃.

Acknowledgment

The author gratefully acknowledges the Egyptian ministry of agriculture which provided me by the Jatropha fruits. I want to thank and recognize the Egyptian ministry of higher education which granted a full Ph.D. scholarship and for Egypt-Japan University of Science and Technology (EJUST) and Japan International Cooperation Agency (JICA) for providing all the facilities to accomplish this work. In addition, I gratefully acknowledge all members of Department of Chemicals and Petrochemicals Engineering Lab for help and support.

References

1. M.F. Demirbas, M. Balat, Recent advances on the production and utilization trends of bio-fuels: a global perspective, Energy Convers Manage, **47**(15), 2371–81 (2006)
2. N.N.A.N. Yusuf, S.K. Kamarudin, Z. Yaakub, Overview on the current trends in biodiesel production, Energy Conversion and Management, **52**(7), 2741–51 (2011)
3. F. Ma, M.A. Hanna, Biodiesel production: a review, Bioresour Technol, **70** (1), 1–15 (1999)
4. S. Mekhilef, S. Siga, R. Saidur, A review on palm oil biodiesel as a source of renewable fuel, Renewable and Sustainable Energy Reviews,;**15**(4), 1937–49 (2011)

5. M.E. Borges, L. Diaz, Recent developments on heterogeneous catalysts for biodiesel production by oil esterification and transesterification reactions: a review, Renewable and Sustainable Energy Reviews, **16**(5), 2839–49 (2012)
6. Y. Zhang, M.A. Dubé, D.D. McLean, M. Kates, Biodiesel production from waste cooking oil: 1. Process design and technological assessment, Bioresource Technol, **89**, 1–16 (2003)
7. Stamenković, O.S., Veličković, A.V., Veljković, V.B., The production of biodiesel from vegetable oils by ethanolysis: current state and perspectives, Fuel **90**, 3141–3155 (2011)
8. M. Berrios, M.A. Martin, A.F. Chica, A. Martin, Study of esterification and transesterification in biodiesel production from used frying oils in a closed system, Chemical Engineering Journal, **160,** 473–479 (2010)
9. Gui MM, Lee KT, Bhatia S., Feasibility of edible oil vs. non-edible oil vs. waste edible oil as biodiesel feedstock, Energy, **33**, 1646–53 (2008)
10. Balat M., Potential alternatives to edible oils for biodiesel production–A review of current work, Energy Convers Manag, **52**(2), 1479–92 (2011)
11. T.M. Yunuskhan, A.E. Atabani, Irfan Anjum Badruddin, Ahmad Badarudin, M.S. Khayoon, S. Triwahyono, Recent scenario and technologies to utilize non-edible oils for biodiesel production, Renewable and Sustainable Energy Reviews, **37**, 840–851 (2014)
12. Leung DYC, Wu X, Leung MKH., A review on biodiesel production using catalyzed transesterification, Applied Energy, **87**:, 1083–95 (2010)
13. Chhetri AB, Tango MS, Budge SM, Watts KC, Islam MR., Non-edible plant oils as new sources for biodiesel production, Int JMol Sci, **9**(2), 169–80 (2008)
14. Koh MY, Mohd Ghazi TI., A review of biodiesel production from Jatropha curcas L oil, Renew Sustain Energy Rev, **15**(5), 2240–51 (2011)
15. Azam MM, Waris A, Nahar NM., Prospects and potential of fatty acid methyl esters of some non-traditional seed oils for use as biodiesel in India, Biomass Bioenergy, **29**, 293–302 (2005)
16. No SY., Inedible vegetable oils and their derivatives for alternative diesel fuels in CI engines: a review, Renew Sustain Energy Rev, **15**, 131–49 (2011)
17. Keith Openshaw, A review of Jatropha curcas: an oil plant of unfulfilled promise, Biomass and Bioenergy, **19,** 1-15 (2000)
18. Janick, Jules; Robert E. Paull (2008). The Encyclopedia of Fruit & Nuts. CABI. pp. 371–372. ISBN 978-0-85199-638-7.
19. Achten WMJ, Mathijs E, Verchot L, Singh VP, Aerts R, Muys B 2007. Jatropha biodiesel fueling sustainability?. Biofuels, Bioproducts and Biorefining **1**(4), 283-291.
20. Achten WMJ, Verchot L, Franken YJ, Mathijs E, Singh VP, Aerts R, Muys B 2008. Jatropha bio-diesel production and use. (a literature review) Biomass and Bioenergy **32**(12), 1063–1084.

THERMAL properties and morphology of Polypropylene/Polycarbonate/Polypropylene-Graft-Maleic anhydride blends

M. S. Mat-Shayuti[1], M. Z. Abdullah[2] and P. S. M. Megat-Yusoff[2]

[1] *Universiti Teknologi MARA, MALAYSIA*
[2] *Universiti Teknologi PETRONAS, MALAYSIA*

Abstract. This work investigates the effect of blending polycarbonate (PC) into polypropylene (PP) matrix polymer on thermal properties and morphology. The blends, containing 5% to 35% of polycarbonate and 5% compatibilizer, were compounded using twin-screw extruder and fabricated into standard tests samples using injection or compression molding. The compatibilizer used was polypropylene-graft-maleic anhydride (PP-g-MA). Thermogravimetric analysis (TGA) showed improved thermal degradation temperature of PP/PC/PP-g-MA blends compared to pure PP. As PC content increased, the thermal degradation temperature also improved. The highest improvement of thermal degradation temperature was 23.3%, demonstrated by 60/35/5 composition. It was found that the thermal stability of PP/PC blends was improved with the addition of PP-g-MA. PP-g-MA was suspected to enhance the phase adhesion between PP and PC, thus improving thermal stability. Microscopy analysis showed PC reinforcement phase existed as particulates dispersed in PP matrix phase. PC also was in irregular shapes of fibers or flakes in certain compositions, depending on PC fraction and compatibilizer content.

1 Introduction

Polymer blending is considered as the most versatile and economic way to engineer a material with specific desired properties. Polymer blend has been expanding widely since the first polymer blending was done between natural rubber with gutta-percha, and currently 36 wt% of total polymer consumption comes from polymer blends, and 80% of plastics are in the form of polymer blends, alloys, and composites [1]. Wide application of polymer blends can be found in areas such as textile, transportation, electronic, and packaging [2, 3].

Polypropylene (PP) is a commodity polymer that has relatively good mechanical properties and low-cost [4]. However, PP is also well-known for having low thermal stability or low heat distortion temperature. This shortcoming of PP makes its applications limited. Food container made from PP could not be used in oven or microwave at a relatively high temperature, electronic parts or semiconductors made from PP has small range of operating temperature, and PP coating in subsea pipelines could not withstand a relatively high crude oil temperature in a long intended lifetime. Thermal properties of PP/polycarbonate (PC) blends had been the subject of few studies and improvement in PP thermal stability was reported [5, 6]. However, little information is available regarding the relation of PC content with thermal properties of PP/PC blends. Furthermore, addition of compatibilizer or third component into ternary PP/PC blends will influence the thermal properties. Therefore, this area of PP/PC blend requires more studies for a clearer and thorough comprehension.

From earlier researches, one of the prime objectives of blending PP with other polymer was to improve PP's disadvantage which is poor impact strength [7, 8]. Blending with PC was attempted to get PP-based material that has considerably good impact strength with sufficient stiffness. Now research on PP/PC blends has moved beyond the impact strength and stiffness, as thermal properties of the blends are also being studied.

PP and PC are immiscible with each other because of differences in polarity and solubility parameters [9], therefore they will exist as separated phases in their blend. The blend will typically possess two glass transition temperatures (Tg), each for PP and PC, and one melting temperature (Tm) of PP. Thermal behavior of PP/PC blends was investigated using differential scanning calorimeter (DSC) [10], and the result showed Tm of PP in the blends were almost similar compared to pure PP. Regardless of the changing ratio of the PP/PC blends, the Tm peaks were almost similar to one another [11]. Tg values were found did not change in PP/PC blend [11].

For the case of immiscible polymer blend, normally compatibilizer is added to increase interfacial adhesion and to reduce interfacial tension between phases in the composite. The addition of compatibilizer does not only influence mechanical properties, but also thermal properties of PP/PC blends. Ning and Ko [12] reported that increasing amount of polypropylene-graft-maleic anhydride (PP-g-MA) from 5 wt% – 20 wt% would lead to higher Tm and lower crytallinity of the blends. In the meanwhile, blends of PP/PC/PP-grafted glycidyl methacrylate (PP-g-GMA) showed that Tm for PP stayed unchanged [13], while the Tg of PC shifted to lower temperature [14]. The addition of PP-g-GMA also resulted in crystallization temperature, Tc, of PP became 5oC – 8oC higher than that in the binary blends of PP/PC. This result might be caused by the PP-g-GMA which acted as a nucleating agent in the PP crystallization process. Dai and Ye [11] found that the Tg peaks of PP shifted to higher temperatures with addition of PP-grafted2-terbutyl-6-(3-tertbutyl-2-hydroxy-5-methylbenzyl)-4-methylphenyl acrylic ester (PP-g-BPA) or styrene-ethylene-butylene-styrene (SEBS), whereas the Tg peaks of PC shifted to lower temperatures.

Renaut and co-workers [6] tested 75% PP, 25% PC, and optional 3% PP-g-MA or Terpolymer ethylene/acrylic ester/maleic anhydride (EBuAMA) blends, and through thermal gravimetry analysis (TGA), it was found that the compatibilized blends showed higher temperature for maximum degradation compared to the noncompatibilized blends. This suggested that thermal stability of PP/PC blends was improved with the addition of compatibilizer. The improvement was attributed to the changes in the morphology of the blends, where adhesion between PP and PC phases might be improved from compatibilizing. Similar reasoning has been reported for other polymer blends [15].

In several occasions, ternary blends of PP/PC with other filler substance have been reported. Chand and Hashmi [5] studied the effect of adding PC into red mud (RM)-filled PP. The thermal stability of PP/PC/RM blends demonstrated improvement over PP/RM, and the higher PC content in the blend, the higher thermal stability achieved. In another research, PP, PC, and attapulgite (AT) were blended together to form PC/PP/AT ternary nanocomposites [16]. The produced blends had higher crystallization temperature of PP, because AT had a strong heterophase nucleation effect on PP. On the other hand, fire testing on intumescent fire retardant of PP/PC/ammonium polyphosphate (APP)/PP-g-MA demonstrated lower heat release rate compared to PP/APP, which is desirable in fire retardant behavior as thermal degradation rate of the materials will be lower [6].

Upon the promising outcomes in the thermal area of PP/PC blends, more research is needed in order to comprehend the matter. Subsequently, this work is aimed to characterize the thermal properties and morphology of PP/PC/PP-g-MA polymer blends of various compositions. PP-g-MA as compatibilizer has been suspected to improve interfacial adhesion between PP and PC phases [6], and is intended in this work to further enhance the thermal stability.

2 Methodology

2.1 Materials

PP used was Propelinas G425, produced by Polypropylene Malaysia, and has melt volume-flow rate (MVR) of 11.0 cm^3/10 min, while PC was from the trade name Panlite® grade L-1225Y, manufactured by Teijin Kasei America Inc. (Teijin Chemicals), with MVR value of 11.0 cm^3/10 min (300°C/1.2 kg). The compatibilizer selected was PP-g-MA from the brand Sigma-Aldrich, with molecular weight of 9100 by gel permeation chromatography (GPC) and maleic anhydride content of 8 – 10%.

2.2 Sample fabrication

PC was dried at 95°C for 12 hours prior to compounding to minimize hydrolytic degradation. Pellets of PP, PC and PP-g-MA were mixed according to designated weight fractions. The blends, containing between 0% to 35% of PC and 5% compatibilizer were compounded by twin-screw extruder at 250°C and 100 rpm before formed into standard shapes by compression molding, with temperature ranging from 190°C to 250°C.

2.3 Microscopy

Scanning electron microscope (SEM) was used to study the morphology of the blends. Specimens were cryogenically fractured and plated with thin gold layer as preparation.

2.4 Thermal characterization

The thermal behaviour of PP, PC, PP-g-MA and their blends were obtained using DSC, Pyris 1, Perkin Elmer Co., USA. The samples were prepared to weight around 0.5 mg, heated from 50°C to 350°C at 10 °C/minute rate alongside a standard reference sample with sample chamber was continuously purged with dry nitrogen. Then the samples were held at 350°C for 5 minutes to remove any thermal history before cooled to 50°C at 10 °C/minute. Graph of heat flow versus temperature then plotted.

For TGA, specimens of about 5 mg were heated from 25°C to 800°C at a heating rate of 10°C/minute. Heating was performed electrically in pure nitrogen gas atmosphere to prevent oxidation and other reaction. Weight changes of the samples were then monitored and tabulated.

3 Results & discussion

3.1 Differential Scanning Calorimetry (DSC) analysis

DSC test was performed to determine the glass transition temperature, T_g and melting temperature, T_m of the blends. The heat flow profile of the DSC test is presented in Fig. 1, and the tabulated data is shown in Table 1. T_m of pure PP was 164.1°C. For PC, its T_g was 156.1°C and it had no T_m, confirming that it was an amorphous polymer. In polymer blend theory, a blend of two immiscible polymers will produce two T_g. However, the T_g for PP could not be determined experimentally, because the theoretical value of T_g for PP is -10°C, which is out of the temperature range for DSC used. All the blends showed almost the same thermal behavior, with very small difference in T_g and T_m values. The T_g of PC values were ranging from 153.8°C – 156.1°C. Finally, the T_m values for all the blends were from 162.0°C – 163.6°C.

For the compatibilized blends, it can be observed from the result that the values of T_g for PC showed reduction, where the blends demonstrated lower values than that of pure PC. The reduction of T_g for PC is suggesting that PP and PC have become slightly miscible with each other because of enhanced interaction between PP and PC phase, resulted from the compatibilization effect.

Table 1. Result summary of DSC test.

Blends	T_g of PC (°C)	T_m (°C)
PP	-	163.6
PC	156.1	-
90/5/5	153.8	162.9
80/15/5	154.1	162.4
70/25/5	153.8	162.0
60/35/5	153.8	162.1
70/30	153.8	162.5

When comparing the compatibilized and noncompatibilized blends from Table 1, for T_g of PC, noncompatibilized blend had value that was very close to compatibilized blends, and the disparity could easily be a result of experimental error. The same findings were reported by Zhihui *et al.* [13, 14] and Dai and Ye [11] when PP-g-GMA, PP-g-BPA or SEBS were used as compatibilizer. The T_g of PC shifted to lower temperature with the addition of PP-g-GMA, PP-g-BPA or SEBS [11], [14].

Next, the melting temperatures of the blends showed slight reduction compared to pure PP. This finding is in agreement with previous research done by Chun *et al.* [10], where no significant changes can be seen for PP/PC blends. The T_m of PP/PC blends were almost similar with each other regardless of PC content [11]. The addition of PP-g-GMA also did not affect T_m values for PP/PC blends, as reported by Zhihui *et al.* [14]. In the meanwhile, it can be observed that the heat of fusion of the blends generally decreases as PC content increases. The same findings were reported previously [13, 14]. From the DSC test, there was no improvement of either T_g or T_m found. This finding was expected, as previous researches reported the same results. The results imply that the expected thermal improvement would not be found in the T_g or T_m of PP/PC blends, but rather in their thermal degradation temperature. The explanation on this can be found in thermal gravimetry analysis (TGA) section later.

Crystallization properties such as crystallization temperature, T_c, and crystallinity, Φ, of the blends can be obtained from DSC test. The T_c values were determined from DSC cooling thermograms, and the values are shown in Table 2. T_c of pure PP was 118.6°C, which is lower than T_g of pure PC which was 156.1°C. This suggests that during cooling process of PP/PC blend, the PC component solidifies before PP begins to crystallize. This behavior will affect the final properties of the blend.

Table 2. Result summary of T_c.

Blends	Crystallization Temperature, T_c (°C)
PP	118.6
90/5/5	122.7
80/15/5	119.9
70/25/5	120.4
60/35/5	118.9
70/30	119.4

All the blends exhibited slightly higher T_c compared to that of pure PP. Differences of the T_c values compared to the pure PP were in the range of 0.3°C – 4.1°C. Similar results were found by Zhihui *et al.* [14], where PP-g-GMA was used as compatibilizer. However, the increment with PP-g-GMA was 5°C – 8°C, which is higher than that of PP-g-MA. From the result, there is no clear indication whether PP-g-MA has an effect towards T_c since the values for compatibilized and noncompatibilized are close to each other. Normally, higher T_c is favorable because it means lower cooling time. Higher T_c may be attributed to PC and PP-g-MA acting as nucleating agents in the PP crystallization process, based on the shifting of crystallization peak temperature for the PP/PC/PP-g-MA blends than that of pure PP. The same reasoning was reported by Renaut *et al.* [6].

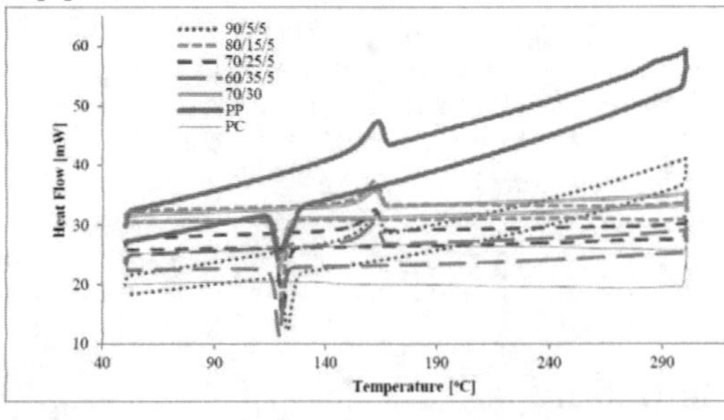

Figure 1. DSC plot of PP, PC, and PP/PC blends of various compositions.

In the meanwhile, crystallinity, Φ were calculated using Equation 1

$$\Phi = \frac{\Delta H_m}{\Delta H_m^0} \times 100\% \qquad (1)$$

where ΔH_m is the value of enthalpy corresponding to the melting of PP and its blends, while ΔH_m^0 is the extrapolated value of enthalpy corresponding to the melting of 100% crystalline PP (ΔH_m^0= 209 J/g). Results of enthalpy of PP melting and crystallinity are summarized in Table 3. For compatibilized blends, the highest degree of crystallization was at 80/15/5 composition, and the lowest is at 90/5/5 composition. The only noncompatibilized blend, 70/30 composition demonstrated the highest crystallinity and it indicates that the crystallinity of PP component in the PP/PC binary blend is higher than that of pure PP. The crystallinity of PP component in PP/PC blends decreases when PP-g-MA is added. It may be deduced that PP-g-MA together with PC will behave as crystallization suppressor in PP matrix. Renaut *et al.* [6] showed that PP/PC/PP-g-MA had smaller degree of crystallization than pure PP, and Ning & Ko [12] reported that increasing amount of PP-g-MA component would lead to lower crystallinity of the blends. These are in agreement with the results obtained in the Table 3. Generally in polymer science, higher crystallinity means stronger material, but at the same time more brittle. Amorphous regions on the other hand contribute towards toughness.

Table 3. Result summary of enthalpy of PP melting, and crystallinity.

Blends	Enthalpy of PP melting, ΔH_m (J/g)	Calculated crystallinity, Φ (%)
PP	67.7	32.4
90/5/5	40.4	19.3
80/15/5	59.1	28.3
70/25/5	51.5	24.6
60/35/5	45.6	21.8
70/30	72.3	34.6

3.2 Thermogravimetry Analysis (TGA)

Degradation temperatures of pure PP, pure PC, and PP/PC blends with and without PP-g-MA were determined using TGA. Results of TGA are depicted in Fig. 2. Results showed that there was improvement in degradation temperature of PP/PC/PP-g-MA compared to pure PP. PP started to decompose at 283°C, PC at 372°C and their blends at 338°C, 340°C, 344°C, 349°C, 333°C for 90/5/5, 80/15/5, 70/25/5, 60/35/5 and 70/30 compositions, respectively. Pure PP, PC, and the blends degraded in one step decomposition mode. Table 4 shows the tabulated date of the TGA test.

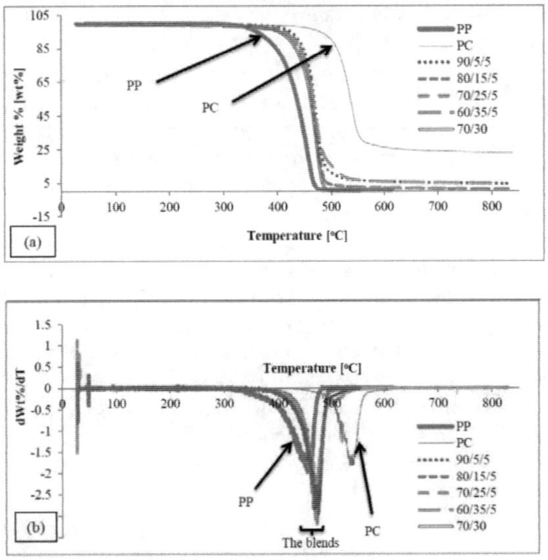

Figure 2. (a) TGA plot, (b) TGA 1st derivative plot.

Table 4. Result summary of thermogravimetry analysis.

Blends	Initial decomposi-tion temp. (°C)	Maximal decomposi-tion temp. (°C)	Final decompos i-tion temp. (°C)
PP	283	455	483
PC	372	539	655
90/5/5	338	470	533
80/15/5	340	472	536
70/25/5	344	472	540
60/35/5	349	473	564
70/30	333	473	492

In general, the blends are presenting intermediate thermal degradation properties between PP and PC. Compared to pure PP, PP/PC blends improvement can be found in initial degradation temperature, maximal decomposition temperature, and final decomposition temperature. The addition of PC improves the thermal stability of PP, and the relationship between thermal stability and PC content is linear. 35% of PC content in 60/35/5 blend composition shows the highest improvement. A study conducted by Chand and Hashmi

[5] reported the same finding where the thermal degradation resistance of PP/PC/red mud blends improved as PC content increased in PP matrix and fixed 5% red mud.

Thermal stability of PP/PC blends is improved with the addition of PP-g-MA. There is a significant enhancement of thermal stability of the compatibilized blend (70/35/5) compared to the uncompatibilized blends (70/30). It may be due to the adhesion between PP and PC phases improved from the compatibilization hence increasing the blends thermal stability, as reported by Renaut *et al.* [6] and Lamas *et al.* [15].

These degradation temperatures values are important for processing purpose. The processing temperature must be high enough for polymers to melt and flow, but it must not exceed the degradation temperature to avoid deterioration of their properties. Finally, it may be deduced that thermal stability of polymer composites depends on two major factors. The first factor is the composition of the constituent polymers, and the second factor is the morphology and bonding between phases in the polymer composites.

3.3 Morphology

The SEM micrographs of the blends after compounding process are shown in Fig. 3, and Table 5 shows the PC particulates diameters for each blend. 90/5/5 composition showed PC particulates dispersed in PP matrix. At 80/15/5 and 70/25/5 compositions, the shape of dispersed PC was a mix of spheres and fibrous. 60/35/5 composition on the other hand showed some PC existed as fibers while the rest had no definite shape, and the size of it was relatively large. Finally, 70/30 composition portrayed PC reinforcement existed as particulates and flakes dispersed in PP matrix.

Table 5. Results of PC particulates diameter from SEM.

Blends	90/5/5	80/15/5	70/25/5	60/35/5	70/30
PC particulates diameter (μm)	3.1 – 4.9	1.9 – 4.4	3.6 – 6.7	–	4.5 – 11.6

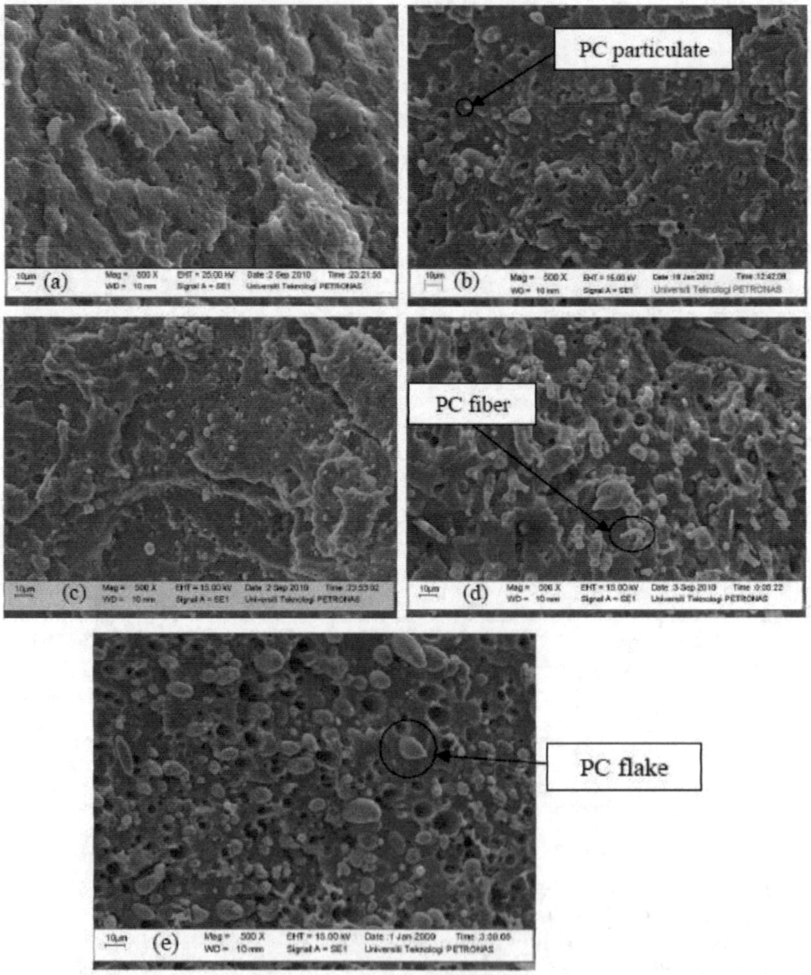

Figure 3. Scanning electron micrograph at 500X magnification of: (a) PP/PC/PP-g-MA (90/5/5), (b) PP/PC/PP-g-MA (80/15/5), (c) PP/PC/PP-g-MA (70/25/5), (d) PP/PC/PP-g-MA (60/35/5), and (e) PP/PC (70/30)

For compatibilized blends of 90/5/5, 80/15/5, and 70/25/5 compositions, the PC particulates diameters were 3.1 µm – 4.9 µm, 1.9 µm – 4.4 µm, and 3.6 µm – 6.7 µm, respectively. For noncompatibilized blend of 70/30 composition, the morphology showed PC particulates diameters between 4.5 µm – 11.6 µm dispersed uniformly in PP matrix.

Generally, it can be concluded that compatibilized blends have smaller diameter of PC particulates compared to the noncompatibilized one [17]. This is due to the compatibilizing effect of PP-g-MA which reduces interfacial tension at PP and PC phase boundary. The reduction in interfacial tension makes them less repulsive towards each other, providing better dispersion and thus smaller PC particulates size [18, 19].

In term of the pattern observed, increasing PC content will lead to an increase in PC particulates diameter. It is suspected that 5% compatibilizer was not sufficient in certain compositions. Zhihui *et al.* [13] reported that for PP/PC blends of 70/30 composition, the ideal compatibilizer amount can reach as high as 20%. Therefore for 60/35/5 system where interfacial area between PP and PC is larger, it is natural to assume that compatibilizer requirement should be higher than that of 70/30 system.

Finally, it can be observed that the color tone of PC reinforcement in compatibilized blends compared to the noncompatibilized blend was different. PC particulates in compatibilized blends appeared to be in a brighter color, while it was darker in noncompatibilized blend. The possible explanation is PP-g-MA coated the outer surface of PC particulate in compatibilized blends. The coating is what giving the slight increase in PP and PC phase adhesion, leading to higher thermal stability.

4 Conclusion

It was found that the addition of PC was successful in improving the thermal stability of PP, shown by the increments in initial degradation temperature, maximal decomposition temperature, and final decomposition temperature. The relationship of thermal stability and PC content is linear. Besides PC content, thermal stability also depends on morphology, and interfacial bonding of the blends. The highest improvement of thermal degradation temperature was 23.3%, demonstrated by 60/35/5 composition.

From the SEM micrographs, it was found that PC reinforcement phase existed as particulates dispersed in PP matrix phase. At 80/15/5 and 70/25/5 compositions, the shape of dispersed PC was a mix of spheres and fibrous. 60/35/5 composition showed some PC existed as fibers while the rest had no definite shape. Finally, 70/30 composition portrayed PC reinforcement existed as particulates and flakes. In general, compatibilized blends have smaller diameter of PC particulates compared to the noncompatibilized one, probably due to the compatibilizing effect of PP-g-MA which reduces interfacial tension at PP and PC phase boundary.

Acknowledgements

This research was under the funding of Research Acculturation Grant Scheme (RAGS/1/2014/TK04/UITM//8) by Ministry of Education Malaysia through Faculty of Chemical Engineering of Universiti Teknologi MARA (UiTM) Selangor, and Mechanical Engineering Faculty of Universiti Teknologi PETRONAS, Malaysia.

References

1. L.A. Utracki, *Polym. Blends Handbook*, Springer Netherlands, pp. 3 – 170 (2014)
2. Y. Xu, Z.D. Sun, X.L. Chen, M. Chen, S.C. Hu, J. Macromolecular Science Part B-Physics **52** pp. 716-725 (2013)
3. E. D. Weil, S. V. Levchik, *Flame Retardants for Plastics and Textiles: Practical Applications* 2nd Edition, Hanser Gardner Pubns, (2016)
4. W. Kaminsky, *Polyolefins: 50 years after Ziegler and Natta I: Polyethylene and Polypropylene (Advances in Polym. Sci.)* Springer, (2013)
5. N. Chand, S. A. R. Hashmi, Bulletin of Material Sci. **22**(4) pp. 801 – 804 (1999)
6. N. Renaut, S. Duquesne, S. Zanardi, P. Bardollet, C. Steil, R. Delobel, J. Macromolecular Sci., Part A: Pure and Appl. Chem. **42** pp. 977 – 991 (2005)
7. B. D. Favis, J. P. Chalifoux, Polym. Eng. & Sci. **27**(21) pp. 1591 – 1600 (1987)
8. K. R. Srinivasan, A. K. Gupta, J. Appl. Polym. Sci. **53**(1) pp. 1 – 17 (1994)
9. N. Nemirovski, A. Siegmann, M. Narkis, J. Macromolecular Sci., Part B **34**(4) pp. 459 – 475 (1995)
10. Y. S. Chun, H. C. Jung, M. S. Han, W. N. Kim, Polym. Eng. & Sci. **39** pp. 2304-2312 (1999)
11. S. Dai and L. Ye, Polym. for Adv. Tech. **19** pp. 1069 – 1076 (2008)
12. P. Ning, T. M. Ko, Polym. Eng. & Sci. **37**(7) pp. 1226 – 1237 (1997)
13. Y. Zhihui, Z. Xiaomin, Z. Yajie, Y. Jinghua, J. Appl. Polym. Sci. **63** pp. 1857 – 1863 (1997)
14. Y. Zhihui, Z. Yajie, Z. Xiaomin, Y. Jinghua, Polym. **39**(3) pp. 547 – 551 (1998)
15. L. Lamas, G. A. Mendez, A.J. Muller, M. Pracella, Eur. Polym. J. **34** pp. 1865-1870 (1998)
16. X. Gao, L. Mao, R. Jin, L. Zhang, M. Tian, Macromolecular Materials and Eng. **290** (9) pp. 899 – 905 (2005)
17. M.S. Mat-Shayuti, M. Z. Abdullah, P. S. M. Megat-Yusoff, *Int. Conf. on Env. and Ind. Innov*. **12** (2011)
18. S. Thomas, Y. Grohens, P. Jyotishkumar, Characterization of Polym. Blends: Miscibility, Morphology and Interfaces 1st Edition, Wiley-VCH, (2015)
19. M.S. Mat-Shayuti, M.Z. Abdullah, P.S.M. Megat-Yusoff, Asian J. of Scientific Research **6** (2), pp. 167 (2013)

MATEC Web of Conferences **69**, 03002 (2016) DOI: 10.1051/matecconf/20166903002
ICCPE 2016

Ibuprofen-Amino Acids Co-Crystal Screening Via Co-Grinding Methods

Muhamad Fitri Othman[1], Norasikin Jamburi[1], Nornizar Anuar[1], Syarifah Abd. Rahim[2] and Nurul Hazwani Rohalim[1]

[1]*Faculty of Chemical Engineering, UniversitiTeknologi MARA Malaysia (UiTM), 40450 Shah Alam, Selangor, MALAYSIA*
[2]*Faculty of Chemical and Natural Resources Engineering, Universiti Malaysia Pahang (UMP), LebuhrayaTunRazak, 26300 Gambang,Pahang, MALAYSIA*

Abstract. The importance of pharmaceutical co-crystals now has been recognized in order to improve the research and development in pharmaceutical industries. Low solubility of active pharmaceutical ingredient (API) has led to the growth of new pharmaceutical co-crystals formation as it enhances the physicochemical properties of the API. In this works, preparation of new co-crystal formation between ibuprofen (IBP) with selected amino acid compounds were performed by using dry grinding and liquid assisted grinding (LAG) techniques. Ibuprofen (IBP) was selected as the API meanwhile glycine (GLY), L-alanine (ALA) and L-proline (PRO) were selected as co-crystal former (CCF) agents. The products of IBP-co-former from grinding experiments for the formation of co-crystals were characterized and verified using X-Ray Powder Diffraction (XRPD), Differential Scanning Calorimetry (DSC) and Fourier Transform Infra-Red Spectroscopy (FTIR). The finding reveals that the IBP-PRO co-crystals have successfully formed. For IBP-PRO system, new crystalline peaks from XRPD were recorded at 2θ values of 4.374°, 5.436° and 10.944° from dry grinding technique and 4.41°, 5.436° and 10.962° for liquid assisted grinding (LAG) technique. A new melting point of 257.49 °C was discovered for IBP-ALA indicates the possibility of co-crystals formation. On the other hand, the analysis for IBP-GLY shows that no co-crystals formed in the system.

1 Introduction

Since decades ago and until now there is no exact definition of co-crystal. Although many researchers came by different definition, co-crystal is commonly accepted as a crystal built out of two or more organic components and basically in their pure solid forms, under ambient conditions. The importance of pharmaceutical co-crystal now has been recognized in order to improve the research and development in pharmaceutical industries [1]. For a poorly soluble active pharmaceutical ingredient (API), the co-crystal approaches will improve the API's solubility [2].

Not limited to solubility, co-crystal also helps in improving other physicochemical properties such as stability, dissolution, and bioavailability of poorly water soluble drugs [3]. The properties of drug co-crystal can be modified by forming pharmaceutical co-crystals that usually has two compounds which are single API and a relevant co-crystal former (CCF). Pharmaceutical co-crystal is a new level of pharmaceutical products as it offers a better option in optimizing the API physicochemical properties without changing the chemical composition of the API [4]. Any changes of properties due to the change of molecular composition of either the API or the CCF will affect the function of the drug.

Grinding method was used as a screening process for the possibility of pharmaceutical co-crystal formation. In this paper, the mechanochemical method with grinding in absence of solvent is termed as "dry grinding". Meanwhile the grinding experiment in the presence of solvent is termed as "liquid assisted grinding" (LAG). Grinding method is the most efficient method for co-crystallization screening process compared to solution crystallization method (the most common method) that relies on evaporation and cooling of stoichiometric solution. This is due to the fact that only a little amount of solvent needs to be used so that the co-crystal formation can occur by reaction in a small liquid phase and/or through amorphous phase formation [5], [6]. Ibuprofen was used as an API while amino acid as CCF is believed to improve the solubility of the API by co-crystallization process due to its zwitterionic state. However, it depends on the side chain of the amino acids which can be acidic or basic and strongly modifies the behaviour of the molecules [7].

One of important rule of thumb before preliminary work for co-crystal screening is carried out is to investigate the possibility of proton transfer involves during hydrogen bonding interaction, which might lead to salt formation instead of co-crystal. The investigation of this possibility can be predetermined by measuring the acid ionization constant, pK_a. A common guideline is the 'rule of three' which states that a salt formation can be expected when the difference of pK_a between the acid and the conjugate acid of the base is greater than 3 units [8]. Empirical solution chemistry demonstrates that a pK_a

difference of at least two units (between an acid and a base moiety) is required to form a salt that is stable in water [9]. Meanwhile, the pK_a difference less than 3 units could result in co-crystal formation as the possibility of proton transfer is little. This rule is only a guideline that can be followed before preliminary work is conducted; however this rule is insufficient in verifying co-crystal formation. It is suggested that a crystal engineering approach must be based on complementary intermolecular interaction rather than just a single pK_a value [10].

2 Material and methods

2.1 Materials

Ibuprofen (IBP) (MW: 206.29 g/mol) was obtained from Shasun Pharmaceutical Limited, India (Batch No.: IB 11070248)(SN Grade).The CCFs used were glycine, L-alanine and L-proline. Glycine (GLY) (MW: 75.07 g/mol) was purchased from Sigma-Aldrich with purity of ≥98% while L-alanine (ALA) (MW: 89.09 g/mol) and L-proline (PRO) (MW: 115.13 g/mol) were purchased from Merck KGaA, Darmstadt, Germany with purity of of ≥98%. Fig. 1 shows the molecular structure of API and CCFs used in this research that obtained from Cambridge Structure Database (CSD).

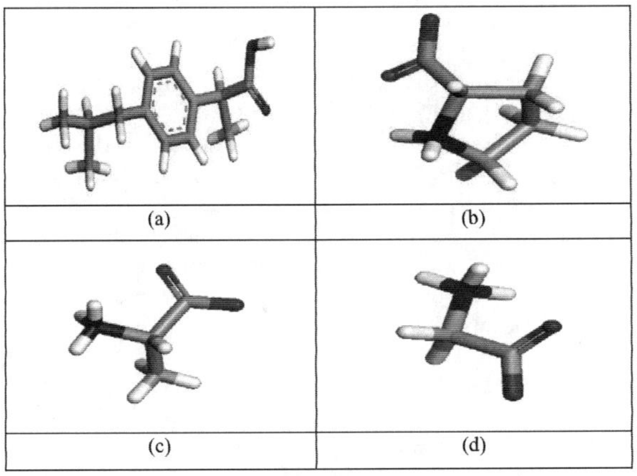

Figure 1. Molecular structure of (a) ibuprofen; (b) L-proline; (c) L-alanine and (d) Glycine obtained from Cambridge Structure Database (CSD).

2.2 Milling methodology

In this work, two milling methods were used; i.e. (dry grinding method and LAG method). Ibuprofen (IBP) was co-milled with one of the selected CCF in oscillatory ball mill (Mixer Mill MM400, Retsch GmbH, Germany) using a 25 mL milling vial. Four stainless steel beads with size of 6 mm in diameter were used in the milling process. The IBP and the CCF were co-milled in stoichiometric ratio of 1:1 (0.01mol, 2.0658g:0.01mol, 1.1541g), 1:2 (0.01mol, 2.0675g: 0.02mol, 2.3378g) and 2:1 (0.02mol, 4.1233g:0.01mol, 1.1553g). The milling process was operated for 30 minutes at frequency of 30 Hz. For LAG method, 2 drops of solvent (ethanol) were added into the mixture of IBP and the CCF before the milling process starts.

3 Characterization methods

3.1 X-Ray Powder Diffraction (XRPD)

The co-crystal formation was determined using an XRPD, Rigaku D/Max-2000 (Rigaku, Japan) with Cu Kα radiation at a wavelength of 1.542 Å in continuous scanning mode. The scanning rate used was 0.1285° over 2θ/min with the step size of 0.018°. The accelerating voltage of the XRPD used was 40 kV and the current was 30 mA. The data were collected over 2θ range of 5° to 30°.

3.2 Differential Scanning Calorimetry (DSC)

Thermal analyses of the products were performed by using DSC (Mettler Toledo DSC 820) equipped with refrigeration cooling system. The products were weighed approximately to 6 mg which then crimped in a non-hermetic aluminium pans (30 μl) and scanned at a heating rate of 10 °C/min in the range of 25 °C to 350 °C. This process was run under continuously purged dry nitrogen environment at a flow rate of 50 mL/min.

3.3 Fourier Transform Infra-Red Spectroscopy (FTIR)

A Thermo Nicolet 6700 FTIR Spectrometer was used to collect the IR spectra of the sample. The information was then used for the analysis of the possibility of proton transfers between the molecules. The scan was collected over the range of 515 to 4000 cm^{-1} for each product. The data was analysed using OMNIC software.

4 Results and discussion

4.1 Ionization constant rule for co-crystals formations

The data was analysed using OMNIC software. The IBP and the amino acids used in this work have more than one pK_a values, reflecting the ionization constants of acidic and basic moieties of the molecules. Table 1 shows the pK_a values of both API and CCFs (i.e. ALA, GLY and PRO) used. Based on the 'rules of three', the result of ΔpK_a values shown in Table 1 indicates that there are possibilities of co-crystals formed instead of salt for IBP-PRO, IBP-GLY and IBP-ALA systems. This is due to the fact that the difference between the pK_a value of the individual component (i.e. between acidic and basic moiety) is more than 2 units, whilst the difference in the pK_a values between the two mixture components is less than 3 units. However the ΔpK_{a2} of IBP-GLY shows the difference of more than 3 units, which might indicate the possibility of non co-crystal to form.

Table 1. The ionization constants for IBP the API and CCFs (ALA, PRO and GLY)

Material	$pK_{a1}{}^*$	$pK_{a2}{}^*$	ΔpK_{a1} (IBP- CCF)	ΔpK_{a2} (IBP-CCF)
Ibuprofen (IBP)	4.45	12.62	-	-
L-alanine (ALA)	2.35	9.69	2.10	2.93
L-proline (PRO)	1.99	10.96	2.46	1.66
Glycine (GLY)	2.34	9.6	2.11	3.02

*The values of $pK_a{}^*$ values were obtained from Dawson, 1959 [11]

4.2 Thermal analysis for preliminary characterization of co-crystal

The screening of the IBP-CCF formation was carried out using two co-grinding methods (dry grinding and LAG). The feed to the co-grinding experiment was 1:1 ratio of IBP to CCF. The ground products from the experiment were analyzed using DSC. Table 2 shows the summary of the result obtained from the experiments.

The result in Table 2 shows the summary of results concluded from the presence of new endotherm peaks detected in the DSC thermogram for the ground samples obtained from both IBP-ALA and IBP-PRO systems. However, for both samples, the presences of pure components were also detected, showing that the samples contain both possible co-crystals compounds and pure components of the starting materials. The dry grinding method was applied to all IBP-CCF mixture system, while the LAG method was only carried out for the IBP-PRO system. The tabulated result shows that the combination of IBP-PRO and IBP-ALA has high possibility of forming co-crystals due to the presence of new endotherm peaks that might corresponds to a new crystalline phase. Meanwhile, the endotherm profile of IBP-GLY shows the presence of two peaks which correspond to pure IBP and GLY components, and with no new peak detected, which could indicate that co-crystal formation did not take place.

Table 2. The ticked signs show the presence of melting peaks of components in the DSC thermogram, representing co-crystals and pure components.

IBP-CC System	Methods	Cocrystals formation	Pure Component
IBP-GLY	Dry Grinding	X	√
IBP-ALA	Dry Grinding	√	√
IBP-PRO	Dry Grinding	√	√
IBP-PRO	Liquid-Assisted Grinding (LAG)	√	√

L-proline appears to be an excellent candidate as a CCF as it shares the zwitterionic α-ammonium-carboxylate synthon, which is common to all other natural amino acids favouring the main chain atom interaction [7]. Different with other amino acids, L-proline is a constrained, rather rigid compound. In terms of formation of co-crystals, this rigidity can certainly be viewed as an entropic advantage over other, more flexible, conformer (entropic contribution) [7]. Assessment of L-proline as a CCF was further investigated using different stoichiometric mol ratio, i.e. 1:2 (IBP:PRO) and 2:1 (IBP:PRO) as both methods show a favourable result for the formation of IBP co-crystal.

The characterization on the thermal behaviour of the samples was analysed by using DSC. Fig. 2(a-d) shows the thermograms patterns for the starting materials (pure API and CCFs) and the ground product of IBP-CCF using selected grinding method.

Thermal behaviour analysis was carried out using the Differential Scanning Calorimetry by heating the samples between 25 °C to 400 °C at 10 °C/min heating rate. The DSC results show the IBP melts at 76.9 °C and evaporates at 285.38 °C. Meanwhile, GLY melts at 253.26 °C, ALA melts at 286.49 °C. From Fig. 2(a), the heat analysis shows that the ground product consists of a mixture of IBP and GLY, and no co-crystal formation detected, as only the pure components peaks were observed. Meanwhile from Fig. 2(b), the endothermic peak of the ground product of IBP-ALA were present at the pure component of IBP peak. However, a new pek was observed at 257.49 °C, which could indicate a new phase. Nevertheless, the presence of two peaks from this results could indicate the presence of IBP and a new phase in the ground product IBP-ALA.

The endothermic peaks of L-Proline shows some variable peaks which indicates the eutectic point and the melting point of the co-former at 118.39 °C and 223.34 °C, respectively [12]. Fig. 2(c) and 2(d) show the results for IBP-PRO systems for both dry grinding and LAG method, using different stoichiometric ratios between the IBP and PRO. For dry grinding method, only one peaks was observed, i.e. at 73.76 °C (for all IBP:PRO ratios), which corresponds for IBP peak. Meanwhile, for LAG method, a new peak was observed at 107.89 °C for 1:1 IBP:PRO ground product, which could indicate a new crystalline phase system. For 1:2 IBP:PRO system, one peak also was observed, at 104.16 °C and this could belong to the PRO peak or a new crystalline phase. However, for 2:1 IBP:PRO system, only IBP peak at 73.76 °C was observed.

The new phase detected from the thermal analysis could belong to the co-crystals, salt crystals or solvates. Thus, a FTIR analysis was carried out to ensure the type of crystals formed in the ground product.

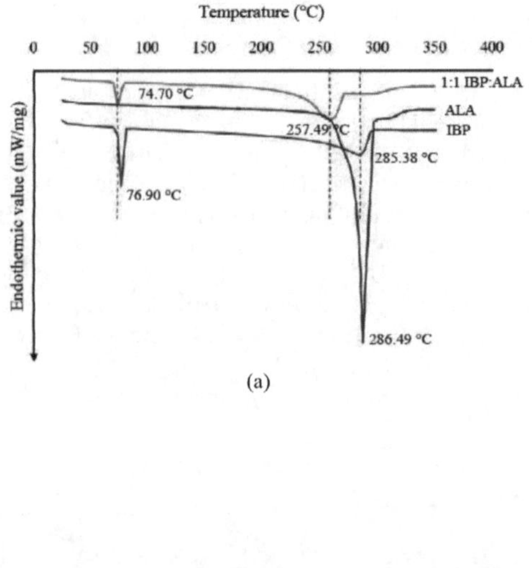

(a)

 DOI: 10.1051/matecconf/20166903002

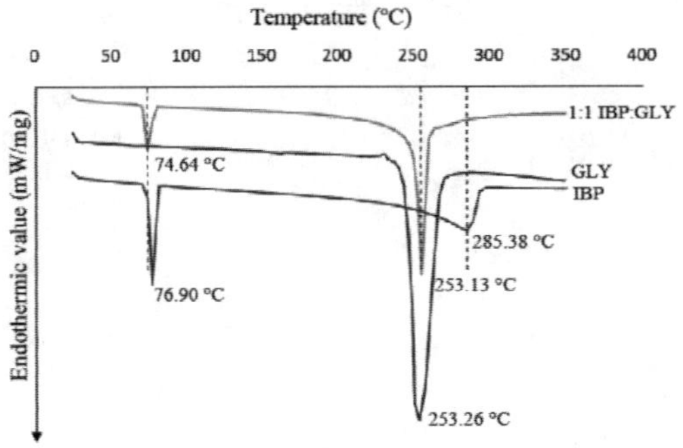

(b)

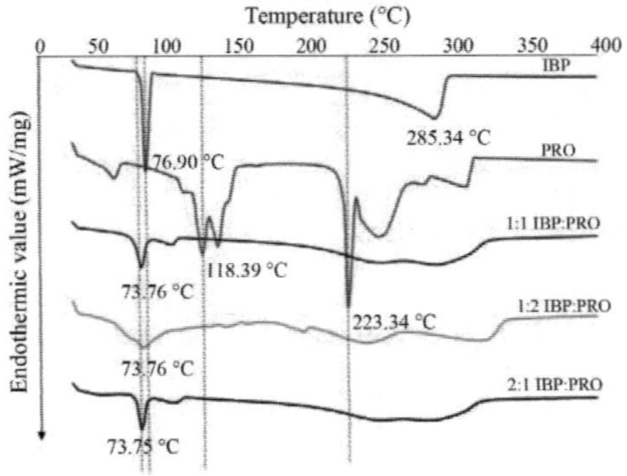

(c)

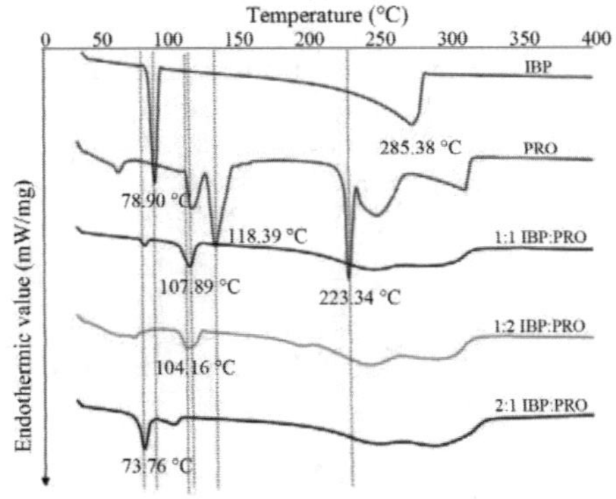

(d)

Figure 2. Thermograms of the sample obtain after grinding of IBP and CCF. (a) IBP and GLY (Dry grinding); (b) IBP and ALA (Dry grinding); (c) IBP and L-Proline (Dry grinding); and (d) IBP and L-Proline (Liquid Assisted Grinding)

4.3 Proton transfer analysis for salt formation

The FTIR spectra of the sample IBP-PRO in the range >1500 cm^{-1} was analysed and the characteristic IR absorption band were compare with the pure IBP and pure PRO system. From Fig. 1 (a), the functional group of ibuprofen consists of aromatic ring and carboxylic acid whilst for L-proline contains an aliphatic ring, carboxyl group and amine group (Fig. 1(b)). The IR spectrum for 1:1 IBP:PRO (LAG) (Fig. 3) shows peaks at 1634.42 cm^{-1} for C=C stretching, 1711.30 cm^{-1} for –C=O in carboxylic acid, 1981.36 cm^{-1} for C-H vibration in R-CH=CH_2, 2553.95 cm^{-1} for –OH in carboxylic acid, 2868.40 cm^{-1} and 2953.81 cm^{-1} for –CH_3. This results show that there are no proton transfer activity for this 1:1 IBP:PRO from LAG method.

There are possibilities of proton transfer at both pure component (IBP and PRO) as both consist of –COOH bond and PRO has more advantage of proton transfer as it has the –NH^+ and –COO^- functional group. The –COO^- of the PRO has the possibility for proton transfer activity by accepting one H^+ to become –COOH and the –NH^+ of PRO will have the possibility for removing one proton to become –NH. Thus from the result, it can be concluded that no solvates were observed, but however, the formation of salt (or proton transfer) for PRO cannot be concluded.

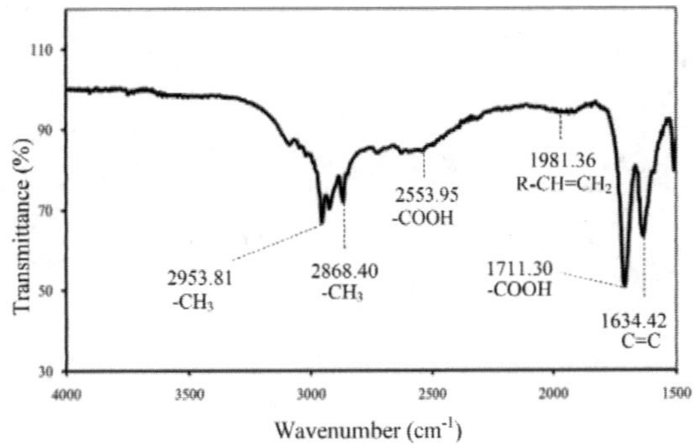

Figure 3. IR Spectra for 1:1 IBP-PRO (LAG method)

4.4 Crystalline structure phase change determination

Fig. 4 (a) and (b) shows the comparison of XRPD patterns of the feed materials of IBP, L-proline and the ground products of IBP-PRO systems recovered from both grinding methods. For the product of IBP-PRO system, ratios between IBP and PRO were varies stoichiometrically, i.e. 1:1, 2:1 and 1:2 for IBP-PRO crystalline products from both grinding.

The XRPD data show that new peaks were detected in all IBP:PRO ratios for both grinding methods. Table 3 also shows that common peaks were recorded for the same stoichiometric ratio for method dry grinding and LAG methods. However, there are some new peaks recorded for LAG samples which were not observed in the product of the dry grinding. This is probably due to the presence of the ethanol solvent in the samples in which ethanol enhances the formation of co-crystal which results in increasing of intensity of new peaks. The same effects also were reported by Trask and co-workers [13] in their work in co-crystallization of caffeine and glutaric acid using dry grinding and LAG methods [14]. For dry grinding method, the XRPD pattern confirms the presence of the unique new crystalline peaks but the DSC results do not show new peak in its thermograms.

By looking at Fig. 4(a) and 4(b) for IBP:PRO ratios of (2:1) and (1:2); peaks for pure components were also detected. For instance, for IBP:PRO ratios of (2:1), IBP peaks present together with the new phase peaks. Whilst for (1:2) ratios, PRO peaks were also detected. This is related to the act of the compound as limiting reactant in the combination. As for IBP:PRO (2:1), the L-proline act as the limiting reactant due to no peak of L-proline discovered but there was peak of IBP discovered for the sample. The remaining of IBP that not consumed lead to the peak of IBP observed in XRPD for the sample IBP:PRO (2:1) ratios in Fig. 4(a) and 4(b). Meanwhile for IBP:PRO (1:2) system, the IBP acted as the limiting reactant as the same peak of PRO observed in the sample. This shows that the stoichiometric of 1:1 IBP:PRO system is the best option as both compound (IBP and PRO) were almost completely consumed as no same peak of both compound observed in the sample.

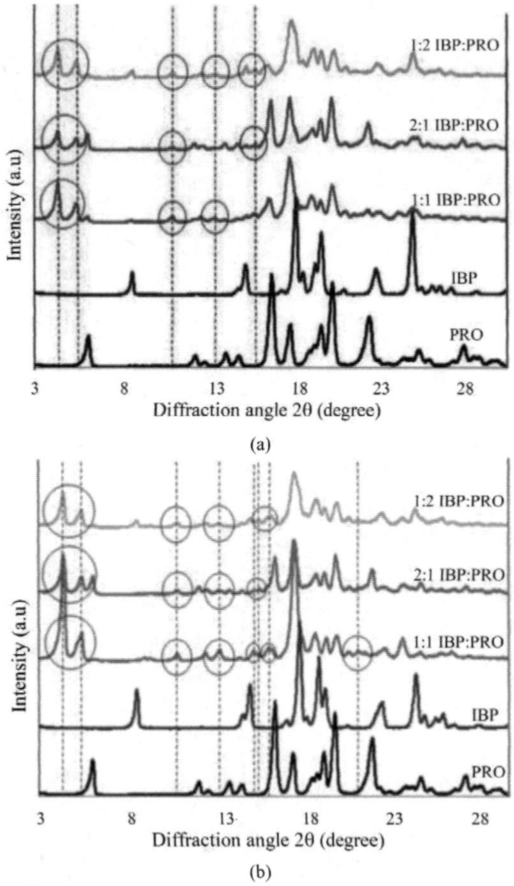

Figure 4. XRPD patterns of IBP, PRO and IBP-PRO ground products obtained from (a) dry grinding method; and (b) Liquid Assisted Grinding method. New peaks were observed and shown by dotted lines on the graph and the 2θ for the new peaks are listed in Table 3.

Table 3. Peaks observed from the XRPD analysis, showing possibility of the presence of a new crystalline form.

Co-crystal Formation Methods	IBP-PRO ratio	New peak	Peak from raw materials
Dry Grinding	1:1	4.338°, 5.382°, 10.854°,13.824°	No peaks from raw materials observed
	2:1	4.32°, 5.364°, 10.818°, 13.30°, 15.606°	Peaks from ibuprofen observed
	1:2	4.338°, 5.382°, 10.872° , 15.57°	Peaks from L-proline observed
Liquid Assisted Grinding (LAG)	1:1	4.392°, 5.436°, 10.926°, 13.338°, 15.300°, 16.182°, 21.438°	No peaks from raw materials observed
	2:1	4.374°, 5.400°, 10.872°, 13.266°, 15.534°, 16.272°	Peaks from ibuprofen observed
	1:2	4.374°, 5.418°, 10.908°, 13.248°, 15.354°, 15.624°	Peaks from L-proline observed

5 Conclusions

The investigation on the possibility of co-crystal formation between ibuprofen (IBP) and three co-crystal formers (CCF) which are L-alanine (ALA), Glycine (GLY) and L-Proline (PRO) were carried out using co-grinding method. There was no co-crystal formation observed on IBP-GLY system and a slight possibility of co-crystal formation for the IBP-ALA system. However, the IBP-PRO system shows high possibility of co-crystal formation and the best combination for IBP-PRO system is by using 1:1 stoichiometric ratio through liquid assisted grinding method in which ethanol was used as a solvent. The other stoichiometric ratio which are 1:2 and 2:1 (IBP-PRO) are not suitable where excess of pure compound were found in the system. The IBP-PRO co-crystal formed was found to obey the 'rule of three' rule in which the pK_a difference is 2.56 units and the new phase was formed, which is suspected a co-crystal. However, salt formation cannot be confirmed.

Acknowledgment

The authors would like to express their gratitude to Universiti Teknologi MARA for funding of this work. This work is part of grants (600-RMI/RAGS 5/3 (211/2014))

References

1. A. V. Trask, Mol. Pharmaceutics, **4**, 9, (2007)
2. S. L. Childs, J. T. Chyall, Dunlap, V. N. Smolenskaya, B. C. Stahly, G. P. Stahly, J. Am. Chem. Soc., **126**, 8 (2014)
3. J. F. Remenar, S. L. Morrissette, M. L. Peterson, B., Moulton, J. M. MacPhee, H. R., Guzma'n, O. Almarsson, J. Am. Chem. Soc., **125**, 2 (2003)
4. William Jones, W.D. Samuel Motherwell, A. V. Trask, MRS Bull., **31**, 4 (2006)
5. R. Thakuria, A. Delori, W. Jones, M. P. Lipert, L. Roy, N. Rodriguez-Hornedo, Int. J. Pharm., **453**, 25 (2013)

6. D.R. Weyna, T. Shattock, P. Vshweshwar, M. J. Zaworotko, Cryst. Growth Des., **9**, 18 (2009)
7. A. Tilborg, B. Norberg, J. Wouters, Eur. J. Med. Chem., **74**, 16 (2014)
8. J. Wouters, L. Quere, RSC, (2012)
9. N. Blagden, M. d. Matas, P. Gavan, P. York, Adv. Drug Deliv. Rev., **59**, 14 (2007)
10. S. L. Childs, G. Stahly, A. Park, Mol. Pharmaceutics, **4**, 15 (2007)
11. R. M. Dawson, *Oxford Clarendon Press*, (1959)
12. A. Tillborg, C. Michaux, B. Norberg, J. Wouters, Eur. J. Med. Chem., **45**, 7 (2010)
13. A. V. Trask, W.D. Samuel Motherwell, William Jones, Chem. Commun., 2 (2014)
14. N. Qiao, M. Li, W. Schlindwein, N. Malek, A. Davies, G. Trappitt, Int. J. Pharm., **419**, 11 (2011)

DOI: 10.1051/matecconf/20166903003

Carbamazepine-Fumaric Acid Co-Crystal Screening Using Solution Based Method

Syarifah Abd Rahim, Chou Choang Tan and Noor Ashila Ramle

Faculty of Chemical & Natural Resources Engineering, Universiti Malaysia Pahang, 26300 Gambang, Pahang, Malaysia

Abstract. Co-crystals is a multi-component system which connected by non-covalent interactions, present physically as a solid form under ambient conditions. Nowadays, co-crystal has becoming as an alternative approach to improve the bioavailability of poor water soluble drugs especially for a weakly ionisable groups or neutral compounds. In this study the co-crystal screening was carried out for carbamazepine (CBZ) and fumaric acid (FUM) co-crystal former (CCF) using non-stoichiometric method (addition of CBZ to CCF saturated solution) and stoichiometric method (evaporation of 1:1 molar ratio of CBZ to CCF) in acetonitrile, ethyl acetate, propanol, ethanol and formic acid solvent systems. The crystals produced from the screening were characterized using Powder X-ray Diffraction (PXRD), Differential Scanning Calorimetry (DSC) and Fourier Transform Infrared (FT-IR). The PXRD analysis had confirmed that the co-crystal was successfully formed in both methods for all of the solvent system studied with an exception to formic acid in the stoichiometric method where no crystal was found precipitate. The findings from this study revealed that Form A and Form B of CBZ-FUM co-crystal had been successfully formed from different solvent systems.

1 Introduction

In general, developers and regulatory authorities in pharmaceutical industry are favouring the pharmaceutical crystal because it provides high purity products that are superior with regard to scalability and reproducibility [1]. The pharmaceutical industry deeply wanted to meet patient's therapeutically needs with the invention of co-crystal where the active pharmaceutical ingredient (API) and co-crystal former (CCF) are playing a key role in formulation development of co-crystal [2].

The pharmaceutical co-crystal can be explained as a multicomponent crystal in which at least one of the molecular components is an API and along with the other component called co-crystal former [3]. The role of co-crystal former is believed to help the active drug to disintegrate into small particles and to be transported to the blood stream where the drug is intended to play its role and still protect the product's stability so that it will be at greatest benefits and effectiveness [4]. It has been reported that the co-crystallisation process allowed the binding of two or more crystal components in single crystalline lattice via hydrogen bonding and van der Walls intermolecular interactions without breaking the bonds or making a new covalent bonds [5], [6].

Co-crystals are considering as a major class of pharmaceutical materials to promote the solubility and dissolution. Apart from co-crystals, other materials such as polymorphs, salts, and amorphous solids are also widely used to enhance dissolution and bioavailability of less soluble API [7], [8]. The enhancement of drug solubility is required in the pharmaceutical industry [9] and solubility is also known as the key factors in determining the efficacy as well as the activity of a drug.

Nowadays, most of the generic drug manufacturers are competing in the increasingly sophisticated market. The improvement of the drug solubility will dramatically draw their attention as this can make profit towards their business [10]. The improvement of the solubility can maximize the bioavailability because poor solubility correlates with the poor bioavailability.

In recent years, carbamazepine (CBZ) has served as a model element for researchers to engage in the study of crystal engineering. CBZ is practically insoluble in water and facing challenges of dissolution-limited bioavailability [3], [11]-[15]. Previous study has reported that co-crystals have been successfully formed using screening methods such as slow evaporation, reaction co-crystallisation, solid state grinding and slurry [16]-[21]. Some examples of the successful CBZ co-crystal are CBZ-nicotinamide co-crystal and CBZ-saccharin co-crystal [6], [16], [18], [22], [23].

The aim of this study is to investigate the CBZ-FUM co-crystal formation using non-stoichiometric method which including continuous shaking and stirring and stoichiometric method of solvent evaporation in five different solvent systems (acetonitrile, ethyl acetate, propanol, ethanol and formic acid).

2 Materials and methods

2.1 Materials

The CBZ and FUM were purchased from ECA International Corporation and Sigma-Adrich Company respectively. Whereas, the solvents (acetone, propanol, ethyl acetate, formic acid, acetonitrile and absolute ethanol) were supplied by Fisher Scientific with purity exceeding 99%.

2.2 Experimental procedures

2.2.1 Non-stoichiometric crystallisation

Saturated solution of the CCF (FUM) was prepared using propanol, ethyl acetate, formic acid, acetonitrile and absolute ethanol in 100 mL conical flask each at room temperature. The solid CCF was added into a fixed amount of solvent. The mixture was agitated at room temperature until the entire CCF was dissolved. More solid CCF was added to the solution manually until some precipitation occurred. The solutions were left to equilibrate for about 72 hours. After 72 hours, the final solutions were filtered from the conical flask using 0.22 μm syringe filter in order to remove excess solid. After filtration, the saturated solution of the CCF was filled in another clean 100 mL conical flask.

The co-crystal was prepared by adding about 0.1-1.2 g of solid CBZ to the prepared saturated solution. The amount of CBZ added varied depending on the type of solvent systems. The solutions were left to equilibrate for about 72 hours in two conditions i.e. in an automated shaker with 150 rpm at room temperature and stirred using a magnetic stirrer. Finally, filter paper was used to separate the final solution slurry and the solid was dried at room temperature. All experiments were performed in triplicate.

2.2.2 Stoichiometric crystallisation

A combination of 1:1 molar ratios of the CBZ and CCF was mixed and stirred to dissolve in a 25 mL of solvent i.e. propanol, ethyl acetate, formic acid, acetonitrile and absolute ethanol in 100 mL conical flask each, at room temperature. More solvent was added until the solutes were fully dissolved in any cases where the complete dissolution was not obtained. Once the solute had fully dissolved, another 10 mL of solvent was added into the final solution. The solution was removed from the conical flask using 0.22 μm syringe filter to remove any impurities. The solution was filled in the 20 mL vial and covered with parafilm, with a few holes poked in in and left to evaporate at room temperature. The produced crystal was filtered using filter paper and dried in the oven at 30^0C for 24 hours. The dried crystal was kept in a vial. All experiments were performed in triplicate.

2.3 Co-crystal characterisations

2.3.1 Powder X-ray diffraction (PXRD)

PXRD was used to determine the presence of co-crystal in the samples by showing different peak profiles. A RIGAKU (Miniflex II) diffractometer was operated at the operating conditions as following: (Cu Kα radiation, voltage 30kV, current 15mA, step size 0.01, step time 1 s and angular range between 3^0 and 40^0 2θ scales.

2.3.2 Differential scanning calorimetry (DSC)

DSC model G1000 with series no. of Q1000-0567 was used to determine the melting point of the co-crystal. The samples (1-3 mg) were crimped in aluminium pans and lid then heated from 30 to 300^0C, under nitrogen purge with flowrate of 50 ml/min at a heating rate of 10 ^{0}C/min.

2.3.3 Fourier transform infrared (FT-IR)

The samples were analysed by FTIR to determine the presence of certain functional groups in a molecule. The analysis was performed using FT-IR with 50 series model attached with the diamond detector at the wave number from 4000-600 cm^{-1} using 32 scans per spectrum with a resolution 4 cm^{-1} for each sample.

3 Results and discussion

3.1 Powder X-ray diffraction (PXRD)

PXRD had been used to identify the presence of co-crystal since every crystalline solid phase has its unique PXRD pattern [24]. The pattern profile shows in Fig. 1 confirmed that CBZ used in this study was Form III [25]. From the analysis obtained from five solvents using stoichiometric and non-stoichiometric method, it was found that co-crystal had been successfully formed with an exception to formic acid in stoichiometry method where no crystal was precipitated. The results were compared with pure components (CBZ and FUM) in order to confirm the formation of co-crystal.

The findings revealed that CBZ-FUM co-crystal was formed in two forms i.e. Form A and Form B depending on the solvent and method used. This finding was agreed with previous, stated that polymorphic transformation of a crystal may affected with the method used during the crystallisation process [21, 26]. According to Childs and co-workers in 2008, they have reported that CBZ-FUM co-crystal Form A and Form B was formed from saturated aqueous and near saturated or saturated ethanolic solution respectively [27].

In addition, the mole ratio of CBZ to FUM used during the experimental work also gave an effect toward the polymorphic form of the co-crystal [28]. The pattern profile of the co-crystal formed from shaking, stirring and evaporation in acetonitrile and ethyl acetate has confirmed that Form A of co-crystal was obtained from both methods of the screening. On the other hand, for formic acid, Form B of co-crystal was formed from only non-stoichiometric method. Different results were found for ethanol and propanol solvent. The form A and Form B co-crystal was formed from stoichiometric and non-stoichiometric method respectively for ethanol whereas, in propanol, Form B was formed from shaking condition (non-stoichiometric) and Form A formed from stirring condition (non-stoichiometric) and stoichiometric methods as shown in Fig. 1.

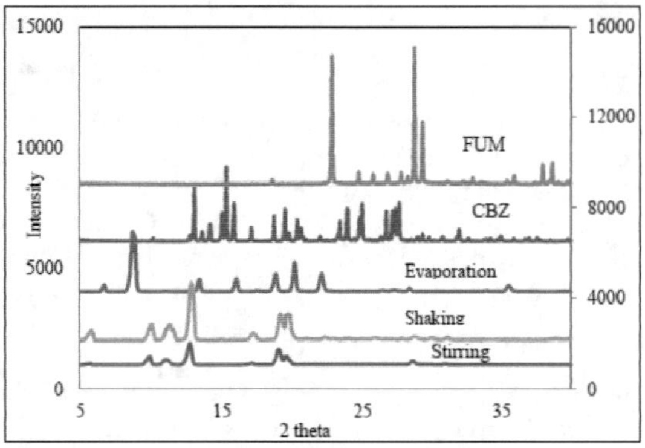

Figure 1. PXRD pattern profile for stoichiometric (form a co-crystal) and non-stoichiometric (form b co-crystal) in ethanol as a solvent.

Table 1. Summary results for stoichiometric and non-stoichiometric method.

Solvent	Screening Method	CBZ-FUM Form A	CBZ-FUM Form B
Acetonitrile	Stirring	√	
	Shaking	√	
	Evaporation	√	
Ethyl acetate	Stirring	√	
	Shaking	√	
	Evaporation	√	
Formic acid	Stirring		√
	Shaking		√
	Evaporation		
Ethanol	Stirring		√
	Shaking		√
	Evaporation	√	
Propanol	Stirring	√	
	Shaking		√
	Evaporation	√	

The overall analysis results including the co-crystals formed from ethyl acetate, formic acid, ethanol and propanol were summarized in Table 1. Table 1 indicated that co-crystals were successfully formed except from solvent evaporation for formic acid which was failed to yield either nor, CBZ, FUM and co-crystal. This is presumably that the solubility of CBZ in the solvent medium directly affected the formation of the CBZ, FUM or co-crystal that required as an optimum condition to induce the precipitation of the crystal in solvent evaporation.

3.2 Differential scanning calorimetry (DSC)

The melting point obtained for FUM was 293^{0}C. On the other hand, there was a weak endothermic peak at 172^{0}C then followed by another endotherm at 191^{0}C for CBZ as shown in Fig. 2. Grzesiak et al. in 2003, has reported that the polymorphic transformation was occurred between 162-175^{0}C and the new phase melted between 189-193^{0}C with a heating rate of 10 ^{0}C/min [25]. This statement has described the polymorphic transformation of Form III to Form I. Therefore the last endothermic peak at 191^{0}C was attributed to the melting point of CBZ Form 1. The finding revealed that DSC analysis is important not only for co-crystal characterization but also for polymorphic characterization, since distinct polymorphs will have different melting temperatures.

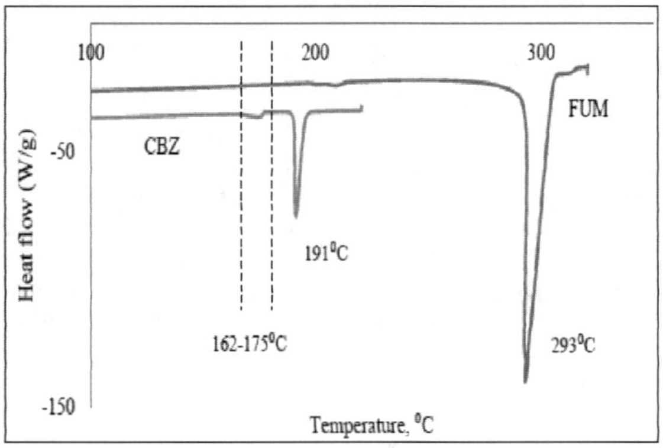

Figure 2. DSC analysis for pure component of CBZ and FUM.

The data obtained from DSC analysis shown that the form A of CBZ-FUM co-crystal has a melting point between 172-176^{0}C [19]. On the other hand, the form B of CBZ-FUM co-crystal has a melting in the range of 186-190^{0}C [21, 27, 28]. The increased of melting point for different of co-crystal polymorphic form was possibly due to the crystal packing nature in the CBZ-FUM co-crystal [1]. The Form A of CBZ-FUM co-crystal has an additional peak appeared in the range of 116-126^{0}C which was due to the existing of water molecule in the co-crystal [27] as shown in Fig. 3.

3.3 Fourier transform infrared (FT-IR)

The peak position in the co-crystal compare with the pure component for acetonitrile was summarized in Table 2. In the FT-IR spectrum of CBZ-FUM co-crystal it shows that the peak profiles was a bit higher or lower frequency compared to the pure component. The significant shift of wavelength had confirmed the formation of the co-crystal [1]. There were some significant peaks presented in the range of 3500 cm^{-1} to 1550 cm^{-1} and 3400 cm^{-1} to 1650 cm^{-1} for the co-crystal in FT-IR analysis results. The occurred peaks were attributed to amides and carboxylic acids functional group of CBZ and FUM.

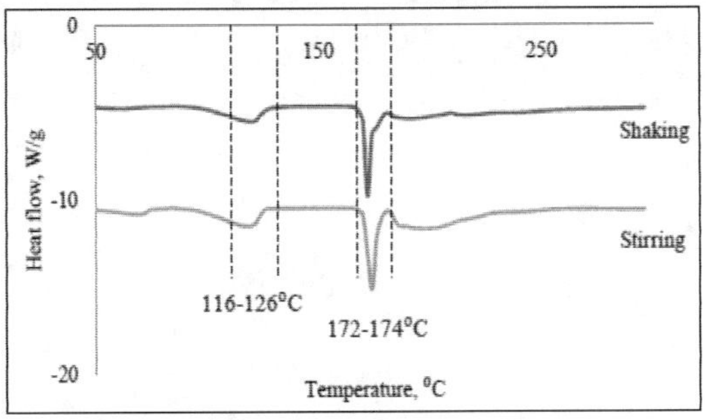

Figure 3. DSC analysis for CBZ-FUM screening obtained from acetonitrile with shaking and stirring (non-stoichiometric method)

Table 2. FTIR data for CBZ, FUM and CBZ-FUM co-crystal based on different screening method in acetonitrile as a solvent.

Type	CBZ	FUM	CBZ-FUM co-crystal		
			Shaking	Stirring	Evaporation
N-H Stretch	3464	-	3455	3455	3455
C=O Stretch	1604	-	1604	1602	1602
N-H Bend	1593	-	1590	1590	1590
O-H Stretch	-	3082	3059	3059	3058
C=O Stretch	-	1659	1660	1659	1656

The FT-IR analysis for ethyl acetate, formic acid, propanol and ethanol had shown the presented of amides and carboxylic acids in the co-crystal but there was some differences between wavelength frequency for both of the functional group in Form A and Form B co-crystal. From the result it shows that Form A with amides and carboxylic acid in range between 3457 cm^{-1} to 1563 cm^{-1} and 3059 cm^{-1} and 1656 cm^{-1} respectively. Whereas for Form B the amides wavelength frequency between 3461 cm^{-1} and 1520 cm^{-1}. The wavelength frequency of 3027 cm^{-1} to1656 cm^{-1} was represented for carboxylic acids.

4 Conclusions

The formation of CBZ-FUM co-crystals were investigated using non-stoichiometric (continuous shaking and stirring conditions) and stoichiometric (solvent evaporation) methods. The PXRD analysis had confirmed that the co-crystal was successfully formed in both methods for all of the solvents systems studied with an exception to formic acid in the stoichiometric method where no crystal was precipitated. From this analysis it was also shows that Form A and Form B co-crystal had been produced from different solvents. This study has revealed that detail screening study is needed for co-crystal formation assessment since the previous and current findings have revealed that different methods, solvent and mole ratio resulted the polymorphic transformation of the co-crystal.

References

1. M. Sevukarrajan, B. Thanuja, R. Sodanapalli, N. Rodríguez-Hornedo, J. Pharm. Sci. Res. (6), 1288 (2011)
2. S.S. Bharate, M.D. Matas, P.T. Gavan, P. York, Expert Opin. Drug Deliv. (9), 1239 (2013)
3. F.L.F. Soares, R.L. Carneiro, J. Pharm. Biomed, Anal. 166(2014)
4. N. Blagden, M. D. Matas, P. T. Gavan, P. York, Adv. Drug Deliv. Rev. (7), 617(2007)
5. A. R. Sonawane, S. S. Rawat, N. N. Janolkar, Asian J. Biomed. Pharm. Sci. (27), 1-8 (2013)
6. A.Y. Sheikh, S. Abd Rahim, R. B. Hammond, K. J. Roberts, Cryst. Eng. Comm. 501-509 (2009)
7. R. Thakuria, A. Delori, W. Jones, M.P. Lipert, L. Roy, N. Rodríguez-Hornedo, Int. J. Pharm. **453**, 101(2013)
8. S.B. Murdande, M.J. Pikal, R.M. Shanker, R.H. Bogner, Pharm. Res. **99**, 1254(2010)
9. S. Kudo,H. Takiyama, J. Cryst. Growth, 87(2014)
10. R.M., Martins, S. Siqueira, L.A. Tacon, L.A.P. Freitas, Powder Technol., 156(2012)
11. R. Nair, S. Gonen, S.W. Hoag, Int. J. Pharm. 11(2002)
12. J.A.H. Junghanns, R.H. Müller, International Journal of Nanomedicine, **3**, 295(2008)
13. M. Moneghini, I. Kikic, D. Voinovich, B. Perissutti, J. Filipović-Grčić, Int. J. Pharm. 129(2001)
14. K. Gowthamarajan, S.K. Singh. Dissolution Technologies, 24(2010)
15. N. Chandel, V. Gupta, A. Pandey, S. Saxena, S. Choudhary, International Journal of Pharmacy & Life Sciences, **2**, 1020(2011)
16. S. G. Fleischman, S. S. Kuduva, J. A. McMahon, B. Moulton, R. D. B. Walsh, N. Rodriguez-Hornedo, M. J. Zaworotko, Cryst. Growth Des. **3**, 909(2003)
17. S.L. Childs, L.J. Chyall, J.T. Dunlap, V.N. Smolenskaya, B.C. Stahly, G.P. Stahly, J. Am. Chem. Soc. **126**, 13335(2004)
18. S.J. Nehm, B. Rodriguez-Spong, N. Rodriguez-Hornedo, Cryst. Growth Des. **6**, 592(2006)
19. A.V. Trask, W.D.S. Motherwell, W. Jones, Int. J. Pharm, **320**, 114(2006)
20. G.G.Z. Zhang, R.F. Henry, T.B. Borchardt, X. Lou, J. Pharm. Sci. **96**, 990(2007)

21. S. Abd Rahim, R.B. Hammond, A.Y. Sheikh, K.J. Roberts, CrystEngComm, 3862(2013)
22. E. Gagnière, D. Mangin, F. Puel, A. Rivoire, O. Monnier, E. Garcia, J. P. Klein, J. Cryst. Growth 311, 2689-2695 (2009)
23. N. Qiao, K. Wang, W. Schlindwein, A. Davies, M. Li, Eur. J. Pharm. Biopharm. *83*(**3**), 415-426 (2013)
24. S. Sehić., G. Betz, Š. Hadžidedić, S.K. El-Arini, H. Leuenberger, Int. J. Pharm. 77(2010)
25. A.L. Grzesiak, M. Lang, K. Kim, and A.J. Matzger, J. Pharm. Sci. **92**, 2260(2003)
26. M. Kitamura, J. Cryst. Growth, 541(1989.)
27. S. Childs and N. Rodríguez-Hornedo, CrystEngComm, **10**, 856(2008)
28. S. Abd Rahim, F. Ab Rahman, E.N.E. Nasir, N.A. Ramle, International Journal of Environmental, Chemical, Ecological, Geological and Geophysical Engineering, **9**, 425(2015)

MATEC Web of Conferences **69**, 03004 (2016) DOI: 10.1051/matecconf/20166903004
ICCPE 2016

A Study on Conceptual Design of Fischer-Tropsch Reactors in GTL Applications

Jae Sun Shin, Faraz Qasim, Jong Hwa Jeong and Sang Jin Park

Department of Chemical Engineering Dongguk University, 26, Pil-dong 3-ga, Jung-gu, Seoul 100-715, Korea

Abstract. GTL (Gas-to-liquid) process is becoming an attractive technology which can produce liquid petroleum products using natural gas. As a part of preliminary design of GTL-FPSO application, process simulation analysis for conceptual design and optimization of reformers and F-T reactors are performed in GTL-FPSO applications by implementing the user made subroutine programs of kinetic equations into PRO/II PROVISION simulator. As for the F-T reactors, Plug Flow Reactor (PFR) model is used with detailed kinetics equations over two different Fe based catalysts (Fe-Cu-K and K/Fe-Cu-Al). Dry reformer is also studied with Plug Flow Reactor (PFR) model. In this study, simulation results are compared with available experimental data and found well agreed with experimental data for both reformer and FT reactor. The Peng-Robinson equation of state is also used to calculate the vapor phase non-idealities and vapor-liquid equilibrium. The optimum operating conditions and process simulation analysis are also presented.

1 Introduction

Due to the era of high oil prices recently, the importance of synthetic oil manufacturing technology using natural gas is emerging gradually and the necessity of an environment-friendly fuel leads to the development of BTL (Biomass-to-liquid), CTL (Coal-to-liquid), and GTL (gas-to-liquid) processes. Among many processes for producing synthetic fuel, GTL process for producing synthetic fuel from natural gas has received much attention for eco-friendly fuel which contains less sulfur and aromatic components [1]-[3]. Since the most of gas fields in the world are small and medium-sized, GTL-FPSO (Floating Production Storage and Offloading) process can be applied for these fields.

Fischer-Tropsch process was developed in 1920s by Franz Fischer and Hans Tropsch and it plays an important role in conversion of synthesis gases in chemical industries [4]-[6].

The application of GTL-FPSO processes has 4 major steps. First step is the production of synthesis gas (syngas) which consists of carbon monoxide and hydrogen gases by reforming reaction [4]-[7]. Second step is the removal process of carbon dioxide and impurities such as water, nitrogen oxides, and hydrocarbons. Carbon dioxide and other impurities should be removed before F-T reactor, which can cause the low efficiency of F-T reactor. This is because the most of stranded gas fields contain lots of carbon dioxide.

In this work, the conceptual design and optimization of reformer and F-T reactors are focused in GTL-FPSO applications using commercial process simulator (PRO/II with PROVISION release 9.1). In reforming process, natural gas is used as a raw material to produce hydrogen and carbon monoxide which results in H_2/CO mole ratio of 2. This ratio is known as an optimum condition for F-T synthetic process to produce synthetic fuel [8, 9].

2 Process description

In dry reforming, there is no water gas shift (WGS) reaction. Only methane and CO2 are reacted to produce H2 and CO as a feed to the F-T process [10]. This reaction can be written in Eq. (1).

$$CO_2 + CH_4 \leftrightarrow 2CO + 2H_2 \qquad \Delta H^0_{300} = \frac{247\text{kJ}}{\text{mol}} \tag{1}$$

As shown in Table 1, operating temperatures and pressures of dry reformer for simulations are changed from 550℃ to 800℃ and 1bar to 30bar, respectively and flowrates of CO2 and CH4 are fixed to 100 kmol/hr while the length of PFR dry reformer varies from 1 to 30 m. A schematic diagram of dry reformer is shown in Fig. 1. In this case, a steady state simulation and optimization are also performed with a plug flow reactor (PFR) and kinetic equations [11] in dry reforming over Ni/La2O3 which is described in Table 2. Subroutine program for these kinetic equations is made and implemented in commercial process simulator (PRO/II with PROVISION release 9.1).

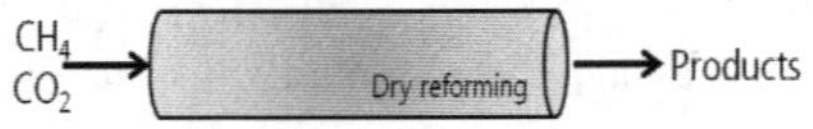

Figure 1. Schematic Diagram of Dry Reformer

Table 1. Feed Conditions to dry reformer

Reactor Length	1~30m	Reactor Temperature	550~800℃
Reactor Diameter	0.0653m	Reactor Pressure	1~30bar
Flowrate (kmol/hr)	CH4	CO2	
	100	100	

Table 2.Kinetic equation and parameters in dry reforming

$$R_{dry} = \frac{K_{1D}k_{2D}K_{3D}k_{4D}P_{CH_4}P_{CO_2}}{K_{1D}k_{2D}k_{3D}P_{CH_4}P_{CO_2}+K_{1D}k_{2D}P_{CH_4}+K_{3D}k_{4D}P_{CO_2}}$$

[mol/(g s kPa)]

$$K_{1D}k_{2D} = 2.61\times10^{-3}\exp\left(-\frac{4300}{T}\right)$$

[mol/(g s kPa)]

$$K_{3D} = 5.17\times10^{-5}\exp\left(\frac{8700}{T}\right)$$

[kPa-1]

$$k_{4D} = 0.531\exp\left(-\frac{7500}{T}\right)$$

[mol/(g s)]

reactor

(Experimental data and kinetic equations are taken from reference [11])

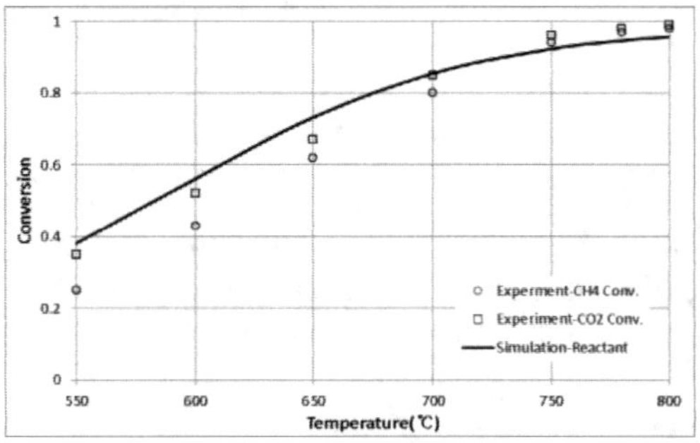

Figure 2. Comparison of experimental data* with simulated conversions of reactants in dry reformer at 1bar and CH4:CO2:He =2:2:6 (*Experimental data are taken from reference [11]).

For the purpose of comparisons, simulated conversions of reactants by using the kinetics equations in Table 2 were based on the same experimental conditions. Experimental

The effect of pressure on the conversion of reactant, for example at T=630, is also analyzed. For operating temperatures lower than 800℃, reactor length should be increased for a given conversion. For the design of commercial scale of reformer and FT reactors, dimension of reactor could be adjusted depending on the amount of feed for processing, operating conditions, the selection of catalysts, temperature and pressure.

In Fischer-Tropsch reaction, the syngas is converting into paraffins and olefins with a wide range of carbon number including waxy components (C20+) over Fe- or Co- based catalysts by chain growth reactions for producing the transportation oil. According to Wang et.al. [12], the most kinetics of FT reactions over Fe based catalysts are lumped into groups which failed to reflect the detailed product information. It is known that Fe based catalyst has several advantages in comparison with Co based catalyst. Fe based catalyst is cheaper than Co based catalyst, and it is less sensitive to impurities in synthesis gas [13-15]. Consequently, the choice of catalysts should be dependent on the economics of the GTL process, the selectivity of product distributions and the operating conditions of the process. In this design of FT reactor, steady state simulations and optimizations are performed with a plug flow reactor (PFR) and two different kinetic equations over Fe-Cu-K (case A) and K/Fe-Cu-Al catalysts (case B) by implementing the user made subroutine programs of these kinetic equations into PRO/II PROVISION simulator and comparisons are reported for the both cases.F-T reactor consist of three major reactions such as the formation of paraffins (CnH2n+2), the formation olefins (CnH2n), and WGS (water gas shift) reactions in the following Eq. (2) to (4).

Paraffin formation

$$nCO + (2n+1)H2 \leftrightarrow CnH2n+2 + nH2O\ (n \geqq 1) \quad (2)$$

Olefin formation

$$nCO + 2nH2 \leftrightarrow CnH2n + nH2O \quad (n \geqq 2) \quad (3)$$

WGS (water gas shift) reaction

$$CO + H2O \leftrightarrow CO2 + H2 \quad (4)$$

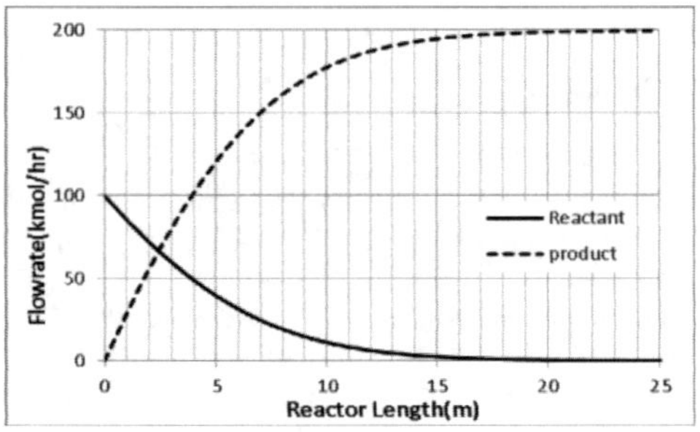

(a)

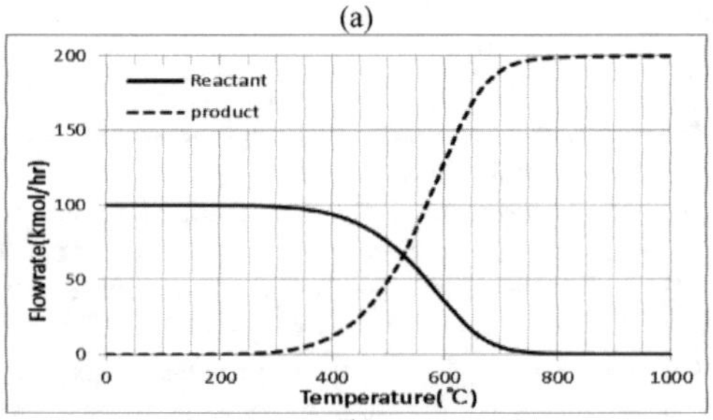

(b)

Figure 3. Flowrates of (a) reactants and products vs. length of reactor at T=800℃ and P=1bar and (b) reactants and products vs. temperature of reactor at P=1bar

3 Results and discussions

Table 3 enlist the feed and reactor conditions for FT reactor. For the application of GTL-FPSO processes, comparisons of experimental with simulated molar flow rates of products are made by using kinetics model over

DOI: 10.1051/matecconf/20166903004

Fe-Cu-K (case A) and K/Fe-Cu-Al (case B) catalysts for F-T reactor wich are shown in 5 and Fig. 6. It should be noted that experimental molar flow rates of products over Fe-Cu-K (case A) catalyst are cumulatively taken with a period of 10-15 h during steady state of FT micro reactor system with the diameter of 18 mm and the length of 300 mm at T=533.15 K, P=20.92 bar for very small amounts of feeds with the ratio of H_2/CO=1.88. For case B, experimental data [17] over K/Fe-Cu-Al catalyst is generated under the conditions with the diameter of 22 mm-i.d. and the length of 1 m at T=523.15 K, P=10 bar for very small amount of feed with the ratio of H_2/CO=1.5, SV=2,000~8,000/h. For the purpose of comparisons, predicted molar flow rates of products by simulations are based on the same operating conditions such as temperature, pressure, feed ratio of H_2/CO, and reactor scale. It is evident from Fig. 4 that the experimental data from the reference agrees well with the simulated data in the present work and there are small deviations between experimental and simulated data.Fig. 5 shows an inverse relation between carbon number and the flow rate. It can be observed that the carbon number get increased if the flow rate is decreased.

Furthermore, the hydrogen to carbon monoxide ratio is about 2 which agrees well with the process requirements. Fig. 6 shows the same behavior for paraffin and olefin. In Fig. 6, the two cases A & B are compared on the basis of flow rate vs carbon number at P=20.92bar, T=523.15K, tube length=10m and diameter of tube=0.0653m. According to Fig. 5 and Fig. 6, all simulated flow rates of products are generally well agreed with experimental data. Comparisons should focus on the trends of product distributions and was shown in similar distribution of products of functional relationship of the carbon numbers. However, though the experimental data for case B seems to be rather inconsistent due to experimental errors, overall comparisons are in acceptable ranges and similar tendency. In Fig. 6, conversions of H_2 and CO increases monotonically as the length of reactor increases for both cases at P=20.92 bar and T=523K. The conversions of CO and H_2 are 0.626 and 0.641 for case A and the conversions of CO and H_2 are shown as 0.720 and 0.756 for case B. Fig. 6 shows that conversions of CO and H_2 for case B are more greater than those for case A.

It is known that existing FT reactors in SASOL processes in South Africa are operating normally in the range of temperature T=350K~523K for LTFT (Low Temperature Fisher-Tropsch) reaction and temperature T=573K~623K for HTFT (High Temperature Fisher-Tropsch) reaction at pressure P=20~21 bar. For the purpose of conceptual design of GTL processes, simulations of FT reactor are performed for two different temperatures T=523K and T=623K at P=20.92 bar even though kinetic parameters for both catalysts A and B are optimized in the range of T=483K~563K. Simulated results for T=623K lie out of the experimental range of temperature but those could be used for interpretation of the effect of HTFT reaction on product distributions from FT reactor.

In this work, simulation results for the FT reactor exhibit that the conversion of the carbon monoxide and hydrogen increases if the operating temperature increases. Same behavior is observed by increasing pressure too. The different combinations of the catalysts are analyzed in this study and the proposed study can lead to a compact and conceptual design of Fischer Tropsch reactors in GTL applications.

Interpretation of the products distribution and selectivity of products can also be done by normalization of products of C_1 to C_5+ excluding the products of CO, H_2O, H_2, and CO_2 assuming the amount of wax formation.

In summary, for both cases A and B, more olefins are produced than paraffins and conversions of CO and H_2 increase as operating temperature and pressure increase which are corresponded to results. In addition, the length of reactor tube should be increased as the diameter of reactor tube decreases for the same conversion of CO and H_2.

Figure 4. Schematic diagram of F-T reactor

Table 2. Feed and reactor conditions to F-T reactor

Flowrate (kmol/hr)		Reactor condition				
CO	H2	Pressure (bar)	Temperature (K)	Length (m)	Tube number	Diameter (m)
100	200	1~30	473 ~1453 .15	1~10 m	5	0.0653m

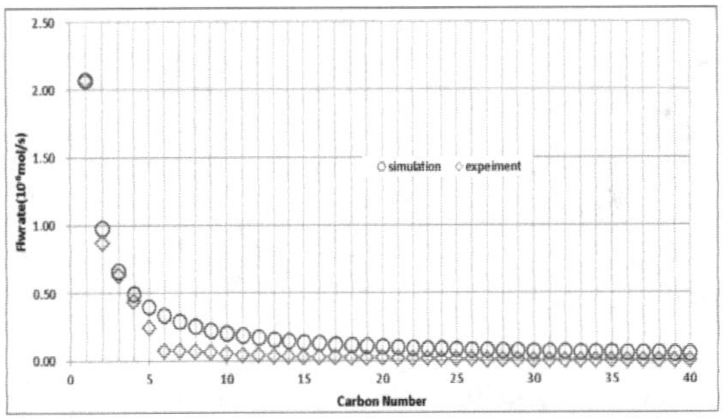

Figure 5. Comparisons of simulation and experimental data* (case A) according to carbon numbers at 533.15K, 20.92bar, and at the feed ratio of H2/CO=1.88 (*Experimental data is taken from reference [16])

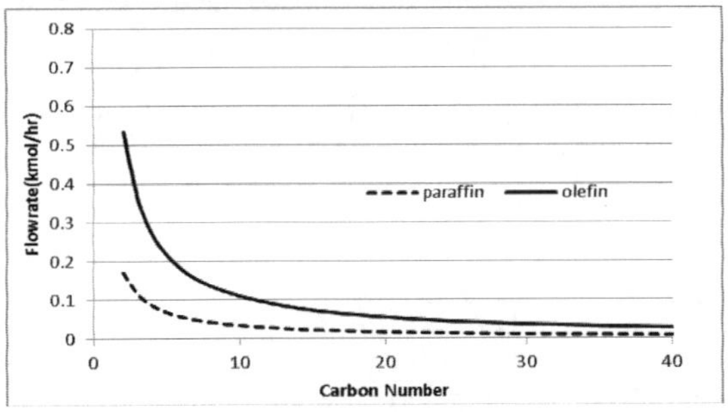

(a)

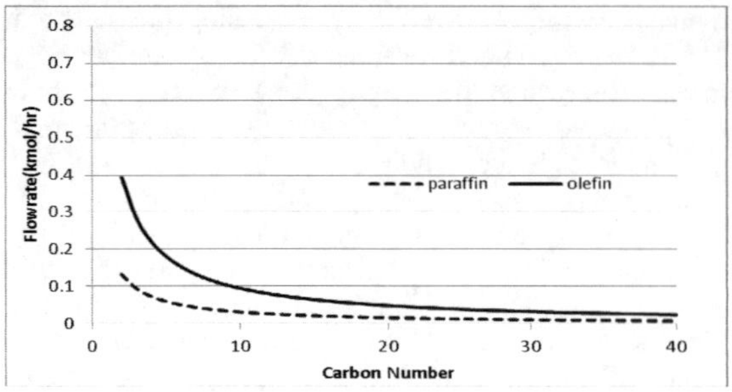

(b)

Figure 6. Comparison of the formation of olefins and paraffins vs. carbon numbers at P=20.92bar, T=523.15K, and tube length=10m and diameter of tube=0.0653m where (a) is for Case A and (b) is for Case B

4 Conclusions

For the purpose of conceptual design of GTL processes, simulations of dry reformer and FT reactor are performed. For FT reactor simulation is done for two different temperatures T=523K and T=623K at P=20.92 bar even though kinetic parameters for both catalysts A and B are optimized in the range of T=483K~563K. Simulation results for FT reactor show that more olefins are produced than paraffins and conversions of CO and H_2 over two different catalysts increase as operating temperature and pressure increase. For the dry reformer, the steady state simulations are performed with exact kinetics equations and optimum operating conditions and reactor scales are reported too. Overall, proposed simulation study can provide the tool for the effective conceptual design of reformer and FT reactor for various catalysts, reactor types and scale, and different operating conditions based on kinetic equations by integrating all GTL processes.

References

1. Y. H. Kim, K. Y. Koo and I. K. Song, Simulation Study on SCR(Steam Carbon Dioxide Reforming Process Optimization for Fischer-Tropsch Synthesis, Korean Chem. Eng. Res., **47**, 700-704 (2009).
2. J. R. Rostrup-Nielsen, Production of Synthesis Gas, Catal. Today, 18, 305-324 (1993).
3. C. Nottenbelt, Mossgas Gas-to-Liquid Diesel Fuels-An Environmentally Friendly Option, Catal. Today, 71, 437-445 (2002).
4. W. Maqbool & E. S. Lee. Syngas Production Process Development and Economic Evaluation for Gas - to - Liquid Applications. Chemical Engineering & Technology, 37, 995-1001 (2014).
5. T. Takeshita, K. Yamaji, Important roles of Fischer–Tropsch synfuels in the global energy future. Energy Policy, 36, 2773–2784 (2008).
6. M. E. Dry, The Fischer-Tropsch Processes, Catal. Today, 71, 227–241 (2002).
7. A.N. Pinheiro, A. Valentini, J.M. Sasaki, A.C. Oliveira, Highly stable dealuminated zeolite support for the production of hydrogen by dry reforming of methane, Appl. Catal. A, 355, 156–168 (2009).
8. D. J. Wilhelm, Simbeck, S. R. Karp, A. D. and Dickenson, R. L., Syngas Production for Gas-to-Liquids Applications: Technologies, Issues and Outlook, Fuel Process Technol., 71, 139-148 (2001).
9. X. Hao, "Simulation Analysis of a Gas-to-Liquid Process Using Aspen Plus", Chem. Eng. Technol., 31, 188–196 (2008).
10. DL Trimm, The formation and removal of coke from nicket catalyst, Catal Rev-Sci Eng. 16, 155-189 (1977).
11. XE Verykio, Catalytic dry reforming of natural gas for the production of chemicals and hydrogen, International Journal of Hydrogen Energy, 28, 1045 – 1063 (2003).
12. J.R. Rostrup-Nielsen, Syngas in perspective, Catal. Today 71, 243–247 (2002).
13. Aasberg-Petersen, K., Bak Hansen, J.-H., Christensen, T. S., Dybkjaer, I., Christensen, P. S., Nielsen, C. S., Winter Madsen, S. E. L. and Rostrup-Nielsen, J. R., Technologies for Large-Scale Gas Conversion, Appl. Catal. A, 221, 379-387 (2001).
14. Y. Yang, H.W. Xiang, R.L. Zhang, B. Zhang, Y.W. Li, A highly active and stable Fe–Mn catalyst for slurry Fischer–Tropsch synthesis, Catalysis Today, 106, 170–175 (2005).
15. A.K. Dalai and B.H. Davis, Fischer-Tropsch synthesis: A review of water effects on the performances of unsupported and supported Co catalysts, Applied Catalysis A: General, 348, 1-15 (2008).
16. Y. J. Lee, S. I. Hong, D, J, Moon, Studies on the steam and CO2 reforming of methane for GTL-FPSO applications, Catalysis Today, 174, 31– 36 (2011).
17. Y. H. Kim, D. Y. Hwang, S. H. Song, S. B. Lee, E. D. Park, and M. J. Park, Kinetic parameter estimation of the Fischer-Tropsch synthesis reaction on K/Fe-Cu-Al catalysts, Korean J. Chem. Eng., 26, 1591-1600 (2009).

DOI: 10.1051/matecconf/20166903005

Variation of Particle Size and Pretreatment Temperature to the Crystallinity of Leucaena Leucocephala

Nor Sharliza Mohd Safaai, Muhammad Izzudin Ibrahim, Amizon Azizan and Habsah Alwi

Universiti Teknologi MARA, Malaysia

Abstract. This study was conducted in order to determine the effect of different particle size and pretreatment temperature to the crystallinity of leucaena leucocephala. The leucaena was pretreated by ionic liquid [1-ethyl-3 methylimidazolium acetate [Emim]Ac. There were three different particle sizes that have been tested in this experiment; less than 0.3 mm, 0.5 mm and 0.7 mm. In the other hand, the pretreatment temperature tested were 30°C, 60°C and 90°C. The effect of particle size and pretreatment temperature to the crystallinity of leucaena was investigated by using X-Ray Diffraction (XRD) and Scanning Electron Microscopy (SEM). The crystallinity index of the sample represents the percentage of crystalline materials. A lower in the cristallinity index indicated that the material has lower crystillinity, hence give more benefit to the cellulose hydrolysis. From XRD analysis, it shows that the cristallinity index of leucaena decreased with decreasing particle size and increasing pretreatment temperature. SEM analysis also shows that the structure of leucaena has more irregular, porous and destroyed structure with decreasing particle size and increasing pretreatment temperature. Thus, the result from this experiment shows that the smaller particle size and higher pretreatment temperature would provide a more accessible surface area to enhance the cellulose hydrolysis.

1 Introduction

An increase in world population and living standards caused the rise of the demand for energy. By 2025, world energy demand is predicted to grow more than 50% due to high demand from several rapidly developing countries [1]. However, the high demand of fossil fuels with the lack of sources has caused the increased of it prices. The researcher found the new liquid fuels that can replace fossil fuels called biofuels. Unlike fossil fuels, biofuels are produced from biomass. Biomass is the biological material from living or recently living organism such as wood, waste materials, gases and alcohol fuels [2]. Cellulosic biomass is the structural component of plants including complex sugars that cannot directly be used for fermentation substrates. There are three main components that make up the cell walls that form the tissue structure of cellulosic biomass which are cellulose, hemicellulose and lignin [3].

Cellulose is a linear polymer consists of long chain made up of glucose that linked together by β-1,4 glycosidic bonds. Cellulose chains are linked together by strong hydrogen bonds that make cellulose highly crystalline polymer [4]. Hemicellulose is an amorphous chain of polysaccharides containing sugar residues and it protects the cellulose fibrils from enzymatic deconstruction [5]. Lignin without sugar-based structure is an amorphous polymer made up of aromatic derivatives [6].

The production of biofuels (Bioethanol) from biomass must go through four main process which are pretreatment, enzyme hydrolysis, fermentation and distillation. The complex structure of lignocellulose biomass cause it difficult to be hydrolyzed and leads to decrease the sugar yield in cellulose hydrolysis. Pretreatment is the most important step in helping enzyme hydrolysis and fermentation steps to maximize the desired product. The goals of pretreatment process are to open up the biomass structure, increase accessible surface area, reduce the cellulose crystallinity and increase the sugar yield in cellulose hydrolysis [7].

Pretreatment are usually classified into four categories which are physical, chemical, physicochemical and biological. It was found that, physical and chemical pretreatment become a conventional pretreatment technologies among the four pretreatment methods [8].

Physical pretreatment involve the size reduction of lignocellulose materials by a combination of chipping, grinding and milling process to reduce the cellulose crystallinity. Chemical pretreatment is classified into alkaline, acid, ionic liquids and wet oxidation. Lately, ionic liquids (IL) have received much attention due to its ability to dissolve various types of biomass. A study by [9] reported that a glucose yield over 90% could be obtained after pretreatment of energy cane bagasse by using [EMIM]Ac. In other study by [10] reported that

[EMIM]Ac is the most efficient among six ionic liquid as it significantly enhanced bagasse enzymatic scarification rate and yield. Furthermore, the low vapor pressure of IL makes its more than 99% recoverable, thus reducing the solvent costs.

Crystallinity can be defined as the degree of structural order in a solid. The degree of crystallinity is used to monitor the level of degradation structure of the material. The effect of particle size and pretreatment temperature to the crystallinity of leucaena was investigated by using X-Ray Diffraction (XRD) and Scanning Electron Microscopy (SEM). From the graph by XRD analysis, the percentage of the cristallinity index can be calculated. SEM is used to monitor the surface structure changes of leucaena before and after pretreatment [11].

2 Methodology

2.1 Sample preparation of leucaena

The leucaena stem was grinded by using grinder machine. The grinded leucaena was sieving into three different particle sizes which are less than 0.3 mm, 0.5 mm and 0.7 mm by using sieve shaker.

2.2 Ionic liquid pretreatment

In order to determine the effect of different particle size and pretreatment temperature to the crystallinity of leucaena, three different particle sizes which are less than 0.3 mm, 0.5 mm and 0.7 mm were used. 0.05 g sample of leucaena with particle size 0.5 mm was mixed with 1 ml of 100% ionic liquid ([Emim]Ac) in a 1.5 ml centrifugal tube. The solution was heated and shacked by using biological shaker at the temperature of 90°C for 3 hour and 1800 rpm shake speed. After cooled at room temperature, the liquid was separated from the precipitated material by using centrifuge at 10000 rpm, 10 minutes and 4°C. The precipitated material was then washed thoroughly with distilled water until no remaining IL solution and dried in an oven at 60°C for 24 hours. The sample was analyzed by using X-Ray Diffraction (XRD) and Scanning Electron Microcopy (SEM). For the different particle size as manipulated variable, the pretreatment was repeated by using different particle size (less than 0.3 mm and 0.7 mm) while as pretreatment temperature become manipulated variable, the pretreatment was repeated by using different temperature (30°C & 60°C).

2.3 Ammonia pretreatment

For ammonia pretreatment, three different particle sizes which are less than 0.3 mm, 0.5 mm and 0.7 mm were used. 0.05 g sample of leucaena with particle size 0.5 mm was mixed with 1 ml of 25% ammonia hydroxide solution in a 1.5 ml centrifugal tube. The solution was heated and shacked by using biological shaker at the temperature of 30°C for 3 hour and 1800 rpm shake speed. After cooled at room temperature, the liquid was separated from the precipitated material by using centrifuge at 10000 rpm, 10 minutes and 4°C. The precipitated material was then washed thoroughly with distilled water until no remaining ammonia solution and dried in an oven at 60°C for 24 hours. The sample was analyzed by using X-Ray Diffraction (XRD) and Scanning Electron Microcopy (SEM).

2.4 Ray diffraction (XRD) and scanning electron microscopy (SEM) analysis

In order to determine the crystallinity index of untreated and pretreated samples of leucaena, the samples were analyzed by using XRD. The samples were analyzed from 4° to 45° with speed angle 2° per minute at 40kV and 40mA. Besides, the morphology of untreated and pretreated samples of leucaena was examined by using SEM. The solid samples were mounted on the aluminum sample stubs with double sided carbon tape. SEM images were respectively taken at 200× magnification with SEM instrument operated at 5kV to 15 kV accelerating voltage.

3 Results and discussion

This experiment was conducted in order to identify the effect of different temperature and particle size on the crystalline of leucaena leucocephala. Three different pretreatment temperature apply in this experiment were 30℃, 60℃ and 90℃ while three different particle size used were less than 0.3 mm, 0.5 mm and 0.7 mm.

In order to understand the effect of different temperature and particle size on the crystalline structure of leucaena, X-Ray Diffraction (XRD) and Scanning Electron Microscopy (SEM) were used for analyzing. XRD analysis was executed to determine the crystallinity of leucaena before and after pretreatment. From the result obtained, the percentage of the crystallinity index could be calculated. In other hand, SEM was utilized to monitor the surface structure changes of leucaena before and after pretreatment.

A lower in the crystallinity index indicated that the material has lower crystillinity. The lower the cristallinity index in leucaena will give benefit to the cellulose hydrolysis.

3.1 Effect of different particle size to the crystallinity of leucaena

Result of XRD analysis for untreated and IL pretreated by using different particle size were shown in Fig. 1. Based on Fig. 1, the scattered intensity at the main peak for cellulose at 2θ (I_{002}) and the minimum intensity between the main and secondary peaks (I_{am}) for each sample may be measured. Then, the crystallinity index for each sample could be calculated by using equation 1.

$$\mathrm{CrI} = \frac{100 \times (\mathrm{I}_{002} - \mathrm{I}_{\mathrm{am}})}{\mathrm{I}_{002}} \quad (1)$$

where, CrI = Crystallinity index (%)

 DOI: 10.1051/matecconf/20166903005

I_{002} = The scattered intensity at the main peak for cellulose

I_{am} = The intensity of the background scatters at 2θ

The percentage of crystallinity index for each sample with different particle size is tabulated in Table 1. It can be concluded that the crystallinity index of Leucaena increased as the particle size increased.

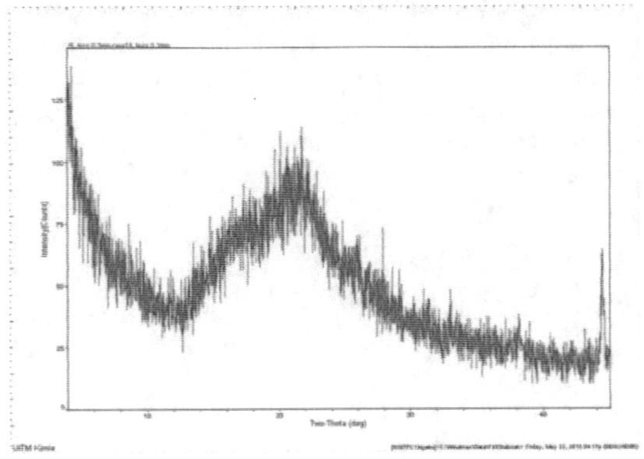

a) Less than 0.3mm

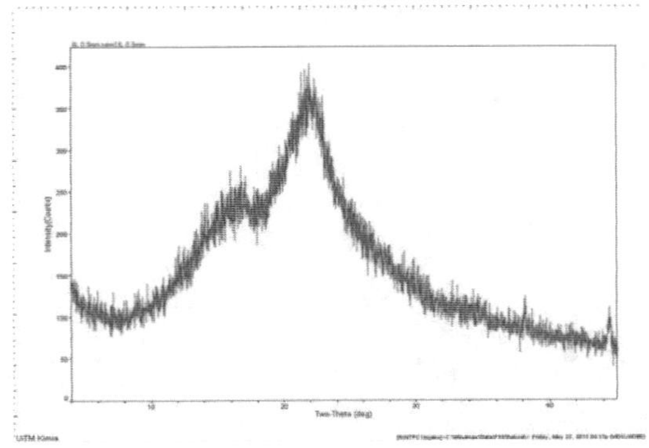

b) 0.5mm

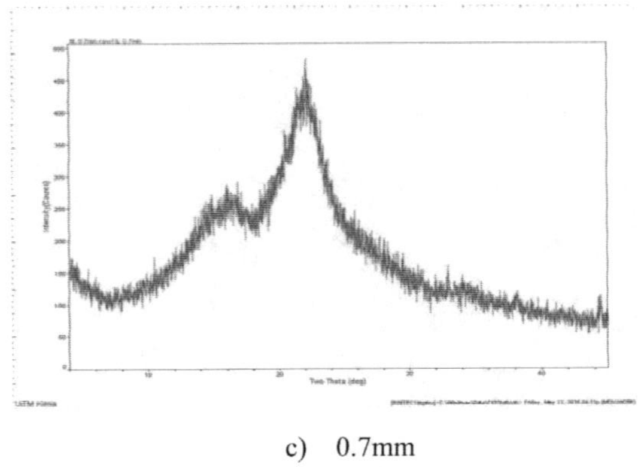

c) 0.7mm

Figure 1. XRD analysis with various particle sizes.

3.1.1 SEM analysis (ionic liquid treatment for different particle size)

Fig. 2 shows the SEM analyzed pictures of untreated and pretreated of leucaena with IL for different particle size. It can be seen that the pretreated leucaena with particle size less than 0.3 mm has a more irregular and destroyed surface compared to the pretreated leucaena with particle size 0.5 mm and 0.7 mm.

Table 1. Effect of particle size on crystallinity index of the leucaena.

Particle Size	CrI (%)
Less 0.3 mm	48
0.5 mm	54
0.7 mm	57

The particle size of sample is one of the crucial factor in the pretreatment because it will directly impact on the contact and diffusion of chemicals into the complex structure of the sample. From Table 1, XRD analysis shows that the cristallinity index of leucaena increased with increased in particle size. This is because decreased in particle size will increase the contact surface area of the sample during pretreatment; hence decrease the crystallinity of the sample.

SEM analysis structure in Fig. 2 is clearly shows that the treated leucaena presents more porous structure compared to the untreated leucaena. The pretreated leucaena with particle size less than 0.3 mm has a more irregular and destroyed surface compared to the pretreated leucaena with particle size 0.5 mm and 0.7 mm. This shows that the smaller the particle size would provide a more accessible surface area to enhance the cellulose hydrolysis.

The result from this experiment shows that smaller particle size will have more advantages since the percentage of glucose conversion will be increased as the particle size of sample decreased.

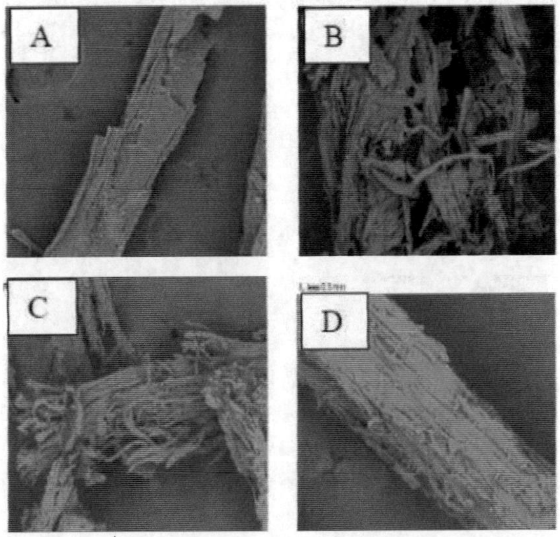

Figure 2. SEM analysis of leucaena (A) untreated, pretreated with IL (B) Less 0.3 mm, (C) 0.5 mm, (D) 0.7 mm.

 DOI: 10.1051/matecconf/20166903005

3.2 Effect of different pretreatment temperature on the crystallinity of leucaena

The crystallinity index by varying the pretreatment temperature was then tabulated in Table 2. It clearly shows that crystallinity index of leucaena decreased as the pretreatment temperature increased. The increased in temperature during pretreatment tend to produce more kinetic energy of the molecule in the sample. Hence, it break down the crystalline structure and decreased the crystallinity of the sample.

Table 2. The effect of temperature on cristallinity index of the leucaena for IL pretreatment.

Temperature (°C)	CrI (%)
30	63
60	56
90	53

3.2.1 SEM analysis (ionic liquid treatment for different temperature)

Fig. 3 shows the SEM analysis while Fig. 4 presents XRD analysis for untreated and pretreated of leucaena with IL for various pretreatment temperature. It can be seen that the increased in pretreatment temperature resulted in more irregular, porous structure and destroyed surface of leucaena. This shows that the higher pretreatment temperature would cause more accessible surface area to enhance the cellulose hydrolysis.

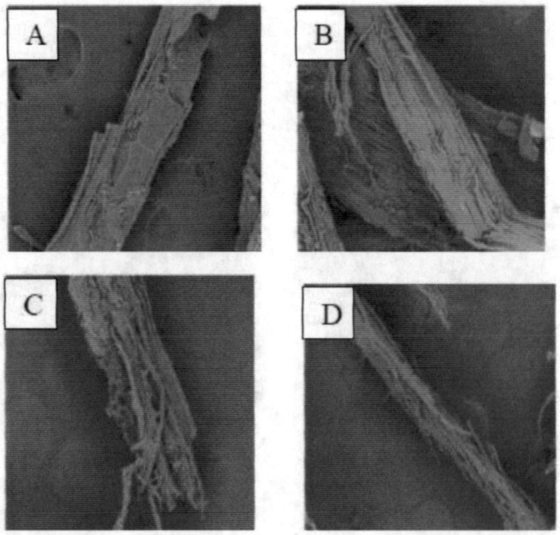

Figure 3. SEM analysis of leucaena (A) Untreated, pretreated with IL (B) 30°C, (C) 60°C, (D) 90°C.

3.3 Comparison of ionic liquid and ammonia pretreatment

IL and ammonia pretreatment shows the same result to the crystallinity of leucaena by using different particle size of sample whereby the crystallinity index for both pretreatment increased as the particle size increased. However, in order to determine which pretreatment method is more efficient between IL and ammonia, the comparison test was conducted.

In this experiment, the particle size, temperature and time were set to constant value. The particle size of the sample used was 0.5 mm, the pretreatment temperature at 30°C and pretreatment time was 3 hours.

Table 3 tabulates the percentage cristallinity index for ionic liquid and ammonia pretreatment. It shows that the crystallinity index of IL pretreatment was lower than ammonia pretreatment. Thus, IL pretreatment is slightly efficient compared to the ammonia pretreatment for leucaena.

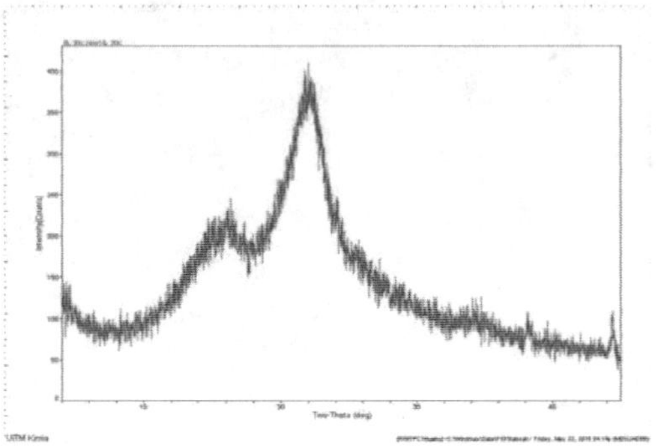

a) 30°C

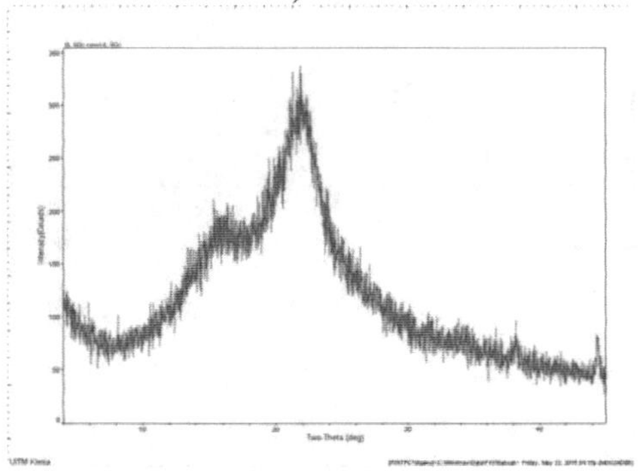

b) 60°C

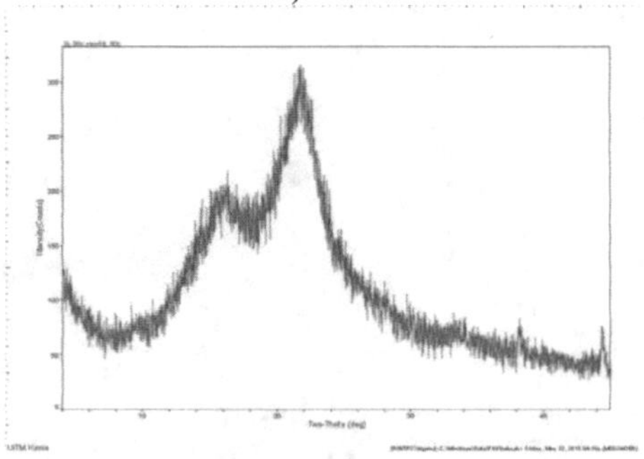

c) 90°C

Figure 4. XRD analysis for untreated and ILtreated lucaena with various pretreatment temperate.

In conclusion, the pretreatment temperature and

particle size of the sample have a significant effect on crystallinity of leucaena. The increased in pretreatment temperature causes the crystallinity index of sample decreased. This is due to the changes in kinetic energy of the molecule whicht also increased. Besides, increased in pretreatment temperature will accelerate the dissolution rates of hydrogen bond in the cellulose. In term of particle size, the crystallinity index of leucaena increased as the particle size increased. It shows that smaller particle size will give more advantages to the cellulose hydrolysis. As the particle size of the sample decreased, the contact surface area and diffusion of IL into the complex structure of the sample will increase

Table 3. The effect of cristallinity index for ionic liquid and ammonia pretreatment.

Pretreatment	Cristallinity Index (%)
IL	63
Ammonia	64

Acknowledgement

This work is supported via financial fund namely the Fundamental Research Grant Scheme (FRGS) 600-RMI/FRGS 5/3 (32/2013) by the Ministry of Higher Education Malaysia and also from Universiti Teknologi MARA, Shah Alam, Malaysia.

References

1. Zhaohui Tong, Pratap Pullammanappallil and Arthur A. Teixeira (2012). How Ethanol is Made from Cellulosic Biomass. University of Florida IFAS Extension. 1-4.
2. Eric M. White (2010). Woody Biomass for Bioenergy and Biofuels in the United States-A Briefing Paper. General Technical Report. PNW-GTR-825. Portland: U. S. Department of Agriculture, Forest service, Pacific Northwest Research Station. 1-45.
3. Yang CY, Fang TJ. Combination of ultrasonic irradiation with ionic liquid pretreatment for cellulase hydrolysis of rice straw. Bioresour Technol 2014;164:198–202.
4. Dadi AP, Schall CA, Varanasi S. Mitigation of cellulose recalcitrance to cellulose hydrolysis by ionic liquid pretreatment. Appl Biochem Biotechnol 2007;137– 140(1–12):407–21.
5. Jacobs A, Palm M, Zacchi G, Dahlman O. Isolation and characterization of water-soluble hemicelluloses from flax shive. Carbohyd Res 2003;338:1869–76.
6. Chen WH, Kuo PC. Torrefaction and co-torrefaction characterization of hemicellulose, cellulose and lignin as well as torrefaction of some basic constituents in biomass. Energy 2011;36(2):803–11.
7. Tan HT, Lee KT. Understanding the impact of ionic liquid pretreatment on biomass and cellulase hydrolysis. Chem Eng J 2012;183:448–58.
8. Sun SN, Li MF, Yuan TQ, Xu F, Sun RC. Effect of ionic liquid/organic solvent pretreatment on the cellulase hydrolysis of corncob for bioethanol production. Ind Crops Prod 2013;43:570–7.
9. Zenghui Qiu, Giovanna M. Aita and Michelle S. Walker (2012) Effect of ionic liquid pretreatment on the chemical composition, structure and enzymatic hydrolysis of energy cane bagasse. Bioresouce Technology 117: 251-256.
10. Aderibigbe S. A., Adetunji O. A. and Odeniyi M. A. (2011) Antimicrobial and Pharmaceutical Properties of The Seed Oil of Leucaena leucocephala (Lam.) De Wit (Leguminosae). African Journal of Biomedical Research 14: 63-68.
11. Francesca Lionetto, Roberta Del Sole, Donato Cannoletta, Giuseppe Vasapollo and Alfonso Maffezzoli (2012) Monitoring Wood Degradation during Weathering by Cellulose Crystallinity. Materials 2012, 5,: 1910-1922.

Effects of particle shape and size on nanofluid properties for potential Enhanced Oil Recovery (EOR)

Tengku Amran Tengku Mohd[1,2], Jumadi Baco[1], Noor Fitrah Abu Bakar[1] and Mohd Zaidi Jaafar[2]

[1]*Faculty of Chemical Engineering, Universiti Teknologi MARA, Shah Alam, MALAYSIA*
[2]*Faculty of Chemical and Energy Engineering, Universiti Teknologi Malaysia, Skudai, MALAYSIA*

Abstract. Application of Enhanced Oil Recovery (EOR) in oil and gas industry is very important to increase oil recovery and prolong the lifetime of a reservoir but it has been very costly and losing properties of EOR agent due to harsh condition. Nanoparticles have been used in EOR application since they are not degradable in reservoir condition and used in smaller amount compared to polymer usage. Commonly, EOR techniques are focusing on increasing the sweep efficiency by controlling the mobility ratio between reservoir fluid and injected fluid. Thus, this research aimed to analyze the nanofluid viscosity at different particle size and shape, volumetric concentration and types of dispersing fluid, as well as to determine the oil recovery performance at different nanofluid concentration. The nanofluid viscosity was investigated at nanoparticle sizes of 15nm and 60nm and shapes of 15nm spherical-solid and porous. Five nanofluid samples with concentration ranging from 0.1wt.% to 7wt.% were used to investigate the effect of volumetric concentration. Distilled water, ethanol, ethylene glycol (EG) and brine were used for the effect of dispersing fluids. Oil recovery was investigated at five different concentrations of nanofluid samples through flooding test. It was found that viscosity of nanofluid increased with decreasing particle size and increasing volumetric concentration. Solid shape particle and increasing dispersing fluid viscosity resulted in higher nanofluid viscosity. The higher the nanofluid concentration, the higher the oil recovery obtained. It can be concluded that nanofluid properties have been significantly affected by the environment and the particle used for potential EOR application.

1 Introduction

Enhanced Oil Recovery (EOR) has become the best solution to solve the problem of declining production and to prolong the life of a reservoir. By using this technique, the recovery performance for a reservoir can be increased up to 60% of oil compared to the conventional method which can only recover between 20-40% of oil [1]. EOR comes with several methods such as injection of miscible solvents and liquid carbon dioxide (CO_2) superfluid, polymer flooding, surfactant flooding, microbial injection and thermal methods. Besides that, another interesting method used is injection of nanofluids [2].

Nanofluid is a mixture of nano-sized solid particles; usually less than 100 μm and its base liquid as shown in Fig. 1. Examples of nanoparticles are metal oxides such as SiO_2, CuO, Al_2O_3, MgO and etc. while for the base fluids are water, brine, ethylene glycol, and etc. Nanofluid can be prepared in several ways such as single-step method and two-step method. One-step method is a process where preparation of nanoparticles and synthesis of nanofluids are done in a two-combined process while two-step method involves synthesizing of nanoparticles first and then mixing with base fluids [3].

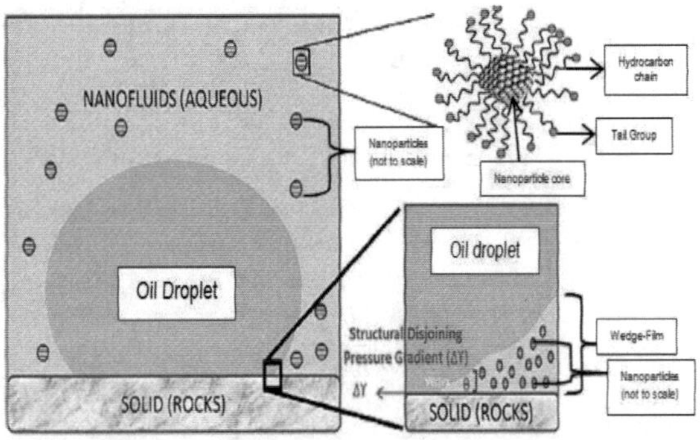

Figure 1. Structure of nanoparticle [3].

For oil and gas industry, one of the mechanisms of nanofluids that can be used for EOR method is viscosity alteration which is very important especially in recovering heavy oil. In addition, it can also be used to control the viscosity of displacing fluid (nanofluid) which then improves the mobility ratio thus increase the oil recovery. Sweep efficiency is influenced by mobility ratio (M) which is also dependent on the mobility of displacing fluid and displaced fluid (oil). The key of controlling the mobility ratio is by altering the viscosity of each one or both of the fluids [4]. This can be expressed in Equation 1.

DOI: 10.1051/matecconf/20166903006

$$M = \frac{\lambda_A}{\lambda_B} \quad (1)$$

where; λ = mobility = k/μ, k = permeability, μ = viscosity, A = displacing fluid and B = displaced fluid. The favorable value for mobility ratio is below than 1 while larger than that indicate low sweep efficiency [5]. The effect of mobility ratio is illustrated in Fig. 2. Basically, low mobility ratio is favorable to displace a fluid whereas high mobility ratio will reduce the sweep efficiency due to the "fingering effect" problem. In the industry, polymers are known to have the ability to increase the viscosity of a fluid but very costly compared to nanoparticles which required only in small amount due to its large surface area [6].

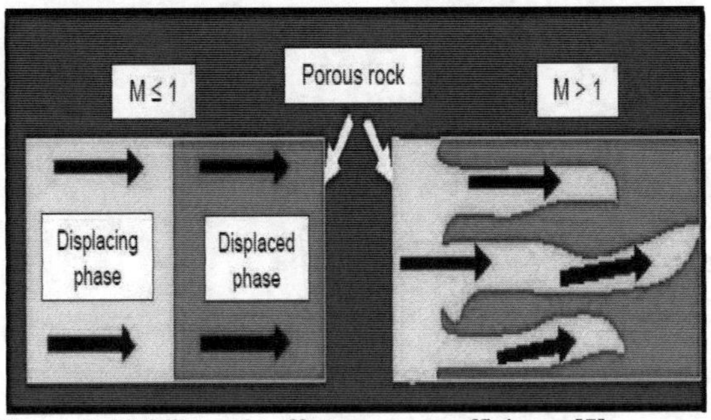

Figure 2. Mobility ratio effect on sweep efficiency [7].

Nanoparticles have been developed for various applications in EOR and can be summarized in three approaches, which are nanocatalysts, nanoemulsions and nanofluids [8]. The presence of nano-sized metal particles catalyzes the breaking of carbon-sulfur bonds within asphaltenes [9]. High viscosity provided by nanoparticle-stabilized emulsions can improve mobility ratio during the flooding, thus increase oil recovery. Mohd et al. (2015) have investigated the mobility of nanoparticle-stabilized CO_2 foam, who found reduction of mobility resulting in enhancing oil recovery [10]. Foam stability also increased with increasing nanoparticle concentration [11]. Zirconium oxide based nanofluids have been applied for wettability alteration, which have great potentials in changing oil-wet limestone towards strongly water-wet condition [12].

Several studies have been done to investigate the particle size effect on the viscosity of a nanofluid. Nguyen et al. (2007) found that the viscosity of nanofluid increased with increasing particle size in their studies on Al_2O_3-water nanofluid [13]. However, several contradictory results have also been obtained. A study by Lu and Fan (2008) on Al_2O_3-water and ethylene glycol (EG) for different nanoparticle sizes showed the decrease in viscosity with increasing particle size [14]. Another study using SiO_2 nanoparticles with sizes of 20, 50, 100nm also found that viscosity reduced with the increase of particle size [15]. There was limited research investigating the effect of particle shape on nanofluid viscosity. However, some studies found that the viscosity of nanofluid was greatly dependent on the particle shape. Timofeeva et al. (2011) reported that elongate particles like platelets and cylinders resulted in higher viscosity compared to spherical particles at the same volume fraction. Other shapes of nanoparticles are spherical, rod, sheet, etc. [16]. Based on previous researches, there was no contradiction about the effect of volumetric concentration on nanofluid viscosity. Das et al. and Putra et al. in 2003 proved that viscosity of nanofluid increased with increasing particle concentration [17, 18]. Thus, this paper aims to analyze the nanofluid viscosity with respect to nanoparticle shape and size, volumetric concentration and types of dispersing fluids, as well as investigating oil recovery at different volumetric concentration.

2 Methodology

Two major experiments have been conducted at room temperature and atmospheric pressure in order to meet the objectives of this research. The first part was the investigation of nanofluid properties and followed by oil recovery flooding test.

2.1 Preparation of nanofluid samples

The main materials used in this research were silicon dioxide (SiO_2) nanoparticles (spherical-solid shape of 15 and 60 nm as well as porous shape nanoparticle) purchased from US Research Nanomaterials Incorporation. These materials were used as they have been received. Paraffin oil, which was used to represent the real oil inside the sand pack for flooding test and Alpha-olefin Sulfonate (AOS) surfactant were provided by Chem Laboratory. Basically, the preparation of nanofluid for this research used two-step method, whereby nanofluids have been formed by dispersing solid nanoparticles into base fluid. Although aggregation of particles was the main challenge for this method, the problem has been minimized by adding some surfactant and application of ultrasonic. Viscosity tests were carried out using Brookfield DV-II+ Viscometer due to its suitability to measure the nanofluid viscosity. Ten readings were taken for every 30 seconds and the average of viscosity was calculated.

2.1.1 Effect of particle size

The nanofluid samples were prepared using SiO_2 nanoparticle. Two different sizes of SiO_2 nanoparticle used were 15 nm and 60 nm; dispersed in distilled water at concentration of 5 wt.%. To increase the nanofluids stability, 0.1 wt.% of surfactant was added and then sonicated for 15 minutes.

2.1.2 Effect of particle shape

The nanofluid samples were prepared using two different shapes of SiO_2 nanoparticle; spherical-solid shape and porous shape. Both samples were dispersed in distilled water at 5 wt.% concentration. 0.1 wt.% of surfactant was added and then sonicated for 15 minutes to increase the samples stability.

 DOI: 10.1051/matecconf/20166903006

2.1.3 Effect of volumetric concentration

SiO_2 nanoparticle was dispersed in distilled water. Five samples of nanofluids were prepared with concentration of 0.1 wt.%, 0.5 wt.%, 2 wt.%, 5 wt.% and 7 wt.%. To retain fluid stability, 0.1 wt.% surfactant was added into each sample and then sonicated for 15 minutes.

2.1.4 Effect of types of dispersing fluid

Four samples of nanofluids were prepared using SiO_2 nanoparticles at concentration of 5 wt.%. Distilled water (DI), ethanol, ethylene glycol (EG) and also brine solution (30000ppm) were used as dispersing fluids with 0.1 wt.% surfactant and then sonicated for 15 minutes to prevent aggregation of particles.

2.2 Oil recovery flooding test

The flooding test was carried out to study the mechanism of nanofluid to recover oil in porous media. Sand pack was used as the porous media while paraffin oil to substitute crude oil. Paraffin oil was coloured with red dye to differentiate it with brine. The volume of oil recovered was divided to the total volume of oil inside the sand pack to measure the percentage of oil recovery. The oil recovery flooding test setup is shown in Fig. 3.

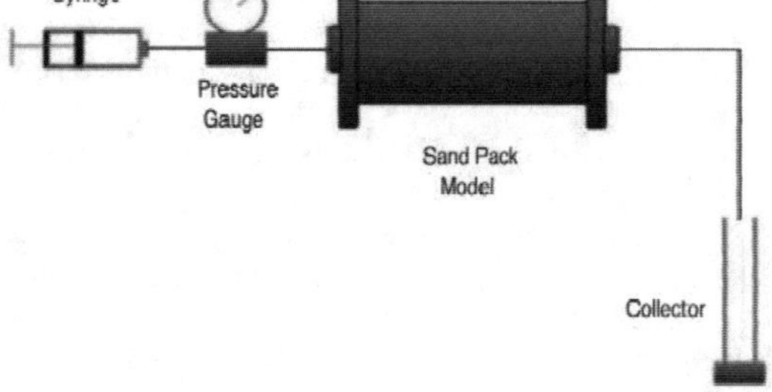

Figure 3. Flooding Test setup.

Sand pack was prepared using sieve shaker to get the uniform size of sand grain between 150-300 microns. The meshed sand was filled into the sand pack holder fully and tightly. The flooding setup includes the main components of sand pack, syringe pump, pressure gauge and measuring cylinder (collector). Firstly, the sand pack was fully injected with brine. Then, paraffin oil was injected into the sand pack until no water can be displaced anymore or called as irreducible water saturation. After that, the secondary recovery was applied to the system using brine which was injected at constant rate of 4 ml/min to recover the oil inside the sand pack. The oil recovered was collected in measuring cylinder and then calculated. Nanofluid was then injected into the sand pack to determine the recovery of EOR. For this purpose, different concentrations of nanofluids were injected at 0.1 wt.%, 0.5 wt.%, 3 wt.%, 5 wt.% and 7 wt.%. The injection was kept at constant rate of 4ml/min and the collected oil recovery in the measuring cylinder was calculated.

3 Results and discussion

3.1 Investigation of nanofluid viscosity

3.1.1 Effect of particle size

Fig. 4 shows the result of nanofluid viscosity versus particle size where the viscosity decreased as the particle size increased.

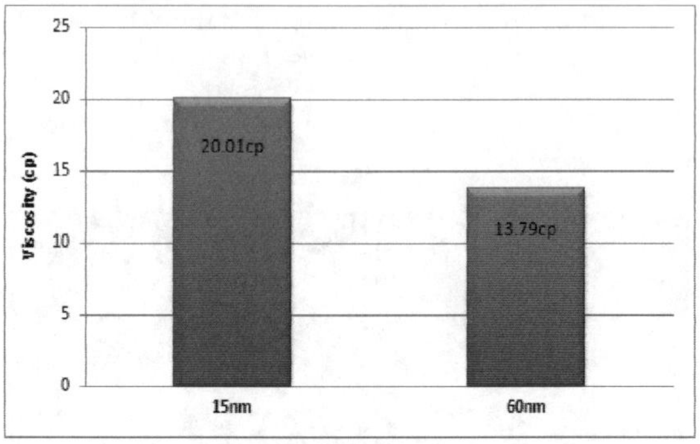

Figure 4. Viscosity versus particle size.

This was supported by finding from Chevalier et al. (2007), who measured the viscosity of SiO_2-ethanol for three different sizes of particle diameter 35, 94 and 190 nm [19]. Similar result was also found by Lu and Fan (2008) and Pastoriza-Gallego et al. (2011), who investigated the viscosity of CuO-water with sizes ranging 23–37nm [14]. According to Bryn McDonagh (2011), in a constant volume fraction, the number of particles increased when particle size decreased resulting in increasing interactions between particles, which then leading to an overall increase resistance to flow [20].

3.1.2 Effect of particle shape

Fig. 5 illustrates nanofluid viscosity versus particle shape, whereby spherical-solid shape provided higher nanofluid viscosity than the sample with porous shape.

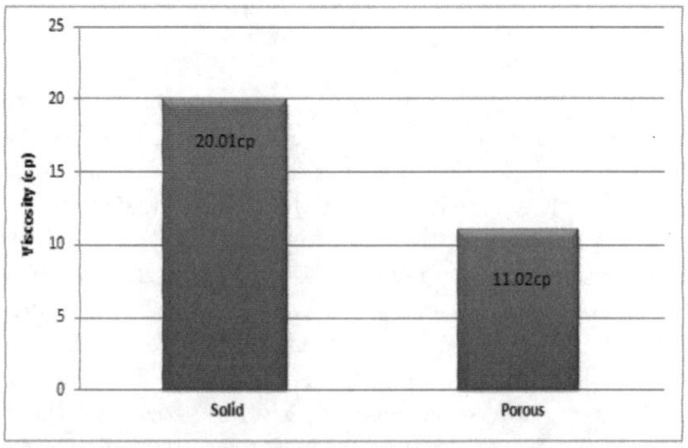

Figure 5. Viscosity versus particle shape.

According to Timofeeva et al. (2009), elongate particle such as platlets and cylinders resulted in high nanofluid viscosity compared to spherical shape [16]. As porosity reduced the resistance of fluid to flow during in contact with surface, hence the viscosity was reduced too [21, 22]. Based on this statement, it can be concluded that the result was acceptable as the viscosity for sample with porous shape was lower than the sample with spherical-solid shape. Although there were not many researches focusing on particle shape, this parameter can be a strong factor in manipulating the viscosity of nanofluid to suit the requirement of its application.

3.1.3 Effect of volumetric concentration

From Fig. 6, it can be clearly seen that the viscosity of nanofluid increased as the volumetric concentration increased. In lower concentration, low amount of particles were inside the water, thus less resistance for the fluid to flow resulting the flow property of water to be dominating. In contrast, for higher concentration solution, there were higher amount of particles, which resulted in more resistance as the particles collided each other and reducing the original flow property of the water; higher viscosity [23]. This result was supported by several researchers, who found identical results. For example, Das et al. (2008) reported that increasing Al_2O_3-water nanofluid concentration has resulted in increasing nanofluid viscosity [24]. Nguyen et al. (2007) concluded that particle size effects are more significant for high particle volume percentage. Their studies on Al_2O_3-water nanofluid exhibited that for volume concentrations below 4%, viscosities corresponding to 36 nm and 47 nm particle size were almost equal. In contrast, for higher concentrations, viscosities were very different [13].

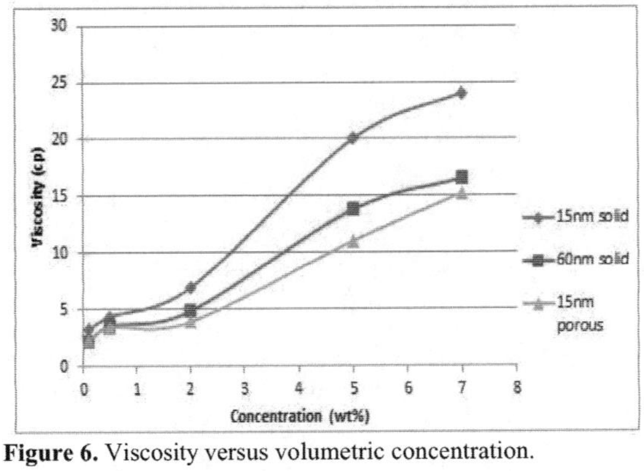

Figure 6. Viscosity versus volumetric concentration.

3.1.4 Effect of dispersing fluid type

Fig. 7 shows the average viscosity of nanofluid samples dispersed in four different fluids. The average viscosities were found to be 20cp, 36.3cp, 19.52cp and 20.43cp for the samples using DI, EG, ethanol and brine, respectively. The highest viscosity was achieved by using EG as the dispersing fluid while the lowest was when using water. These results were affected by the viscosity of the dispersing fluid itself as the viscosity of EG and ethanol were approximately 20cp and 1cp, respectively at room temperature. Thus, it can be concluded that the viscosity of nanofluid was dependent on the viscosity of its dispersing fluid.

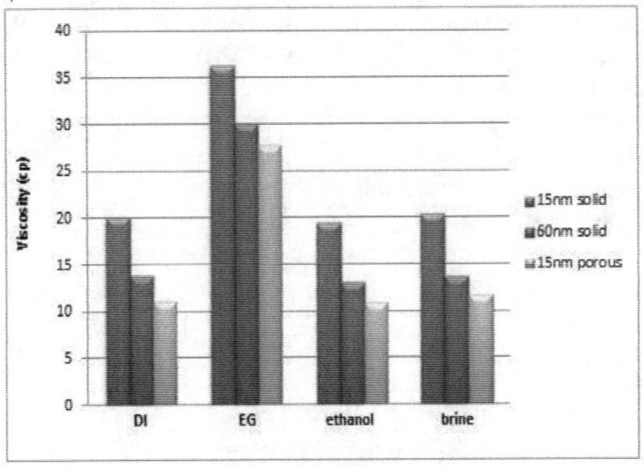

Figure 7. Viscosity versus type of dispersing fluid.

The average viscosity for samples using DI and brine were almost the same to the sample using water as the original viscosities for the three dispersing fluid were nearly the same. The investigation of nanofluid viscosity with respect to different dispersing fluid is important as the property of dispersing fluid could be one of the factors to influence the nanofluid produced. Besides that, further studies of dispersing fluid could be beneficial to enhance the application of nanofluid. Selection of dispersion fluid used was based on several researches such as in a study of Al_2O_3-water and ethylene glycol (EG) for different volume fractions and in a study of SiO2-ethanol with three different nanoparticle sizes [19, 14].

3.2 Investigation of oil recovery

Fig. 8 shows the result of oil recovery based on varying nanofluid concentration. From the figure, the oil recovery increased as the concentration of nanofluid increased.

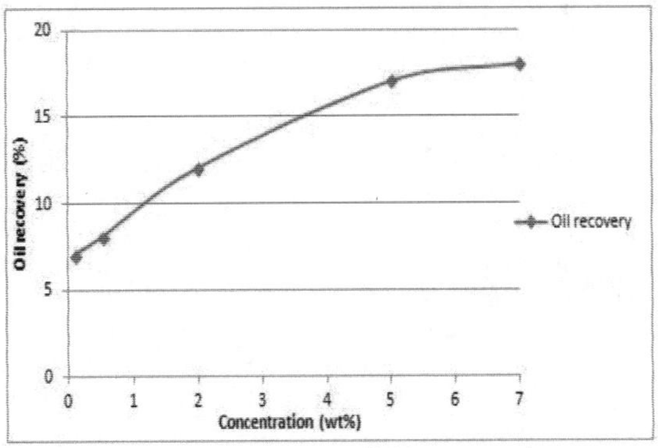

Figure 8. Oil recovery versus nanofluid concentration.

The lowest oil recovery was obtained from the injection of 0.1 wt.% of nanofluid with percentage of 7% recovery while the highest recovery was from the injection of 7 wt.% of nanofluid with 18% recovery. However, the injection of 7 wt.% of nanofluid might contribute to some problems due to higher precipitation of nanoparticle. According to Hendraningrat et al. (2013), high nanoparticle concentration and injection rate could lead to high impairment. The retention of nanoparticles inside the core plugs induced porosity and permeability with some blockage to reservoir hence reduced the porosity [25].

From this experiment, small difference of oil recovery (1%) was found between 7 wt.% and 5 wt.% of nanofluid. Furthermore, the precipitation rate for 5 wt.% sample was low, indicating small effect on the reservoir porosity. By this consideration, it can be concluded that 5wt% nanofluid was the most effective concentration to be used in recovering oil as it could provide high oil recovery, low precipitation rate and less impact to the reservoir properties. For this flooding test, only SiO_2 nanoparticle with size of 15nm and spherical-solid in shape was chosen in preparing the nanofluid samples. The selection was based on previous viscosity tests, which resulted in highest viscosity and DI as the dispersing fluid by considering the economic benefit for industrial application.

This result also proved the Equation 1 which theoretically explained that high viscosity of displacing fluid reduced the mobility of that fluid, resulting in mobility ratio less than one [4]. Fig. 8 shows the highest percentage of oil recovered was from the highest concentration of nanofluid sample, and vice versa. This was due to 7 wt.% nanofluid concentration had the highest viscosity which could recover the paraffin oil inside the sand pack more effectively. Maintaining low mobility ratio is very important to increase the sweep efficiency for recovering more oil and minimizing 'fingering effect' [5, 7].

4 Conclusion

The present study has revealed that the application of EOR can help to increase the total oil recovery and prolong the life of reservoir. From this experiment, it can be concluded that viscosity of nanofluid increased as the particle size decreased, volumetric concentration increased, using of solid shape particle instead of porous shape and also using high viscosity of dispersing fluid. From the flooding test, the percentage of oil recovery increased with increasing concentration of injected fluid but too high concentration could damage the reservoir productivity. The outcomes could be very useful for industrial application as it provides new knowledge on nanofluid properties with respect to particle shape and size, volumetric concentration and types of dispersing fluids to fully optimize the potential of nanofluid properties. This could be potentially applied in nanofluid flooding of EOR application.

Acknowledgment

The authors would like to thank the Faculty of Chemical Engineering, UiTM Shah Alam for the facilities and constant encouragement. This work was financially supported by the Fundamental Research Grant Scheme (FRGS): 600-RMI/FRGS 5/3 (28/2013).

References

1. A. Amarnath, "Enhanced oil recovery scoping study," (1999).
2. J.J Taber, F.D. Martin, and R.S Seright, "EOR screening criteria revisited," Paper presented at the Symposium on improved oil recovery (1996).
3. H.T Zhu, Y.S Lin, and Y.S Yin, "A novel one-step chemical method for preparation of copper nanofluids," Journal of Colloid and Interface Science, 277(1), 100-103 (2004).
4. J.G Richardson, and H.L Stone, "A quarter century of progress in the application of reservoir engineering," Journal of Petroleum Technology, 25(12), 1371-1379 (1973).
5. T.D. Ma, and G.K Youngren, "Performance of immiscible water-alternating-gas (IWAG) injection at Kuparuk River Unit, North Slope, Alaska," Paper presented at the SPE Annual Technical Conference and Exhibition (1994).
6. N. Kothari, B. Raina, K. Chandak, V. Iyer, and H. Mahajan, "Application of ferrofluids for enhanced surfactant flooding in IOR," Paper presented at the SPE EUROPEC/EAGE Annual Conference and Exhibition (2010).
7. R.S. Seright, 'Clean Up of Oil Zones after a Gel Treatment," Paper presented at the SPE International Symposium on Oilfield Chemistry (2005).
8. R. A. F. Elsayed. "A Comparative Study between nanoparticles Method and the Other Methods for Increasing Heavy Oil Recovery", MSc. thesis, Suez University (2014).
9. Y. H. Shokrlu and T. Babadagli: "Transportation and Interaction of Nano and Micro Size Metal Particles Injected to Improve Thermal Recovery of Heavy-Oil", SPE 146661-PA, DOI: 10.2118/146661-PA (2013).
10. T. A. T. Mohd, N. Alias, N. A. Ghazali, E. Yahya, A. Sauki, A. Azizi, & N. M. Yusof. "Mobility Investigation of Nanoparticle-Stabilized Carbon Dioxide Foam for Enhanced Oil Recovery (EOR)". Advanced Materials Research, 1119, 90–95 (2015).
11. T. A. T. Mohd, A. H. M. Muhayyidin, N. A. Ghazali, M. Z. Shahruddin, N. Alias, S. Arina and N. A. Ramlee. "Carbon Dioxide (CO_2) Foam Stability Dependence on Nanoparticle Concentration for Enhanced Oil Recovery (EOR)". Applied Mechanics and Materials, 548-549, 1876–1880 (2014).
12. L. N. Nwidee, S. Al-anssari, A. Barifcani, M. Sarmadivaleh & S. Iglauer. "Nanofluids for Enhanced Oil Recovery Processes : Wettability Alteration Experimental Set-Up". Paper presented at the Offshore Technology Conference Asia, Kuala Lumpur, Malaysia, 22–25 March 2016 (2016).
13. C. Nguyen, F. Desgranges, G. Roy, N. Galanis, T. Mare, S. Boucher and H. Anguemintsa, "Temperature and particle-size dependent viscosity data for water-based nanofluids–hysteresis phenomenon," Int. J. Heat Fluid Flow 28 (6) 1492–1506 (2007).
14. W. Lu and Q. Fan, "Study for the particle's scale effect on some thermophysical properties of

nanofluids by a simplified molecular dynamics method," Eng. Anal. Boundary Elem. 32 (4) 282–289 (2008).
15. P.K. Namburu, D.P. Kulkarni, A. Dandekar and D.K. Das, "Experimental investigation of viscosity and specific heat of silicon dioxide nanofluids," Micro Nano Lett. 2 (3) 67–71 (2007).
16. E.V. Timofeeva, W. Yu, D.M. France, D. Singh and J.L. Routbort, "Nanofluids for heat transfer: an engineering approach," Nanoscale Res. Lett. 6 (1) 182 (2011).
17. S.K. Das, N. Putra and W. Roetzel, "Pool boiling characteristics of nano-fluids," Int. J. Heat Mass Transfer 46 (5) 851–86 (2003).
18. N. Putra, W. Roetzel and S.K. Das, "Natural convection of nano-fluids," Heat Mass Transfer 39 (8)775–784 (2008).
19. J. Chevalier, O. Tillement and F. Ayela, "Rheological properties of nanofluids flowing through microchannels," Appl. Phys. Lett. 91 (23) 233103 (2007).
20. B. Mcdonagh, "Wiley-Blackwell discount." Universal, Fortuna (2011).
21. R. E. Larson, and J. J. L. Higdon. "Microscopic flow near the surface of two-dimensional porous media. Part 1. Axial flow." *Journal of Fluid Mechanics* 166: 449-472 (1986).
22. L. Hendraningrat, B. Engeset, S. Suwarno, S. Li and O. Torsæter, "Laboratory investigation of porosity and permeability impairment in Berea sandstones due to hydrophilic nanoparticle retention", Paper SCA2013-062, California, USA (2013).
23. N. Putra, W. Roetzel and S.K. Das, "Natural convection of nano-fluids," Heat Mass Transfer 39 (8)775–784 (2008).
24. S.K Das, S.U. Choi, W. Yu, and T. Pradeep, "Nanofluids: science and technology," Wiley-Interscience Hoboken, NJ (2008).
25. L. Shidong, L. Hendraningrat, and O. Torsaeter. "Improved Oil Recovery by Hydrophilic Silica Nanoparticles Suspension: 2 Phase Flow Experimental Studies." IPTC 2013: International Petroleum Technology Conference (2013).

Theoretical study of two states reactivity of NO activation on iron atom

Xiaoman Wang, Yaxin Su, Wenyu Liao and Wenyi Deng

School of Environmental Science and Engineering, Donghua University, Shanghai 201620, P.R.China

Abstract. The mechanism of Fe+NO was calculated by the Density Functional Theory (DFT) with the B3LYP methods combined with the 6-311+G (d, p) basis set. The geometry of reactants, transition states, intermediates and products of two reaction systems were completely optimized, and all the transition states were verified by the vibration analysis and intrinsic reaction coordinate (IRC) calculations. The "Two State Reactivity (TSR)" was used to analyze the reaction mechanisms; Results showed that the reaction system preferentially involves low-spin state entrance channel and the high-spin state exit channel. In the reaction channel, the crossing point appears, which would effectively reduce the activation energy and increase the release of reaction heat, play a significant and beneficial role in the kinetic and thermodynamic aspects of this catalytic reaction.

1 Introduction

The nitrogen oxides (NO_x) emission from fossil fuel combustion is a major environmental problem we are currently facing and the elimination of NO_x from flue gases emitted from coal-fired boilers and vehicles is an important issue in atmospheric pollution control. The catalytic reduction of NO_x is a very effective technology for the removal of NO_x. Over the last two decades, transition metal clusters have attracted much interest in the application of heterogeneous catalysis due to their unusual physical and chemical behaviour [1-5]. Among the transition metals, iron is a desirable catalyst owing to the abundant reserves, environmental friendly character and active properties, which was investigated by researchers widely. Recent experiments showed that metallic iron can directly reduce NO into N_2 according to the following reaction [6, 7].

$$Fe + NO \rightarrow N_2 + Fe_xO_y \quad (1)$$

Unfortunately, there is not a solid conclusion on the final chemical forms of iron oxides so far. By XRD (X-ray Diffraction)analysis, Su et al. [8] found that the iron surface was oxidized after reducing NO and both FeO, $Fe_3O_4$and Fe_2O_3 were formed depending on the temperature. However, the XRD analysis could only check the final reaction products during NO reduction by iron. The reaction pathway and evolution process, binding, energies, etc., was not fully understood yet.

It has been known that Density Functional Theory (DFT) can offer significant insight to understand the nitric oxide activation or reduction by small iron clusters. For example, Blagojevic and co-workers [9] investigated the catalytic decomposition of nitric oxide by the Fe^+ .The results show that Fe^+ can lower the activation energy effectively. Heyden et al. [10] researched nitrous oxide decomposition over Fe-ZSM-5 in the presence of nitric oxide by means of DFT. They reported that single iron sites ($Z^-[FeO]^+$) are converted into a novel active Iron center ($Z^-[FeOH]^+$) in the presence of NO, these sites are capable of promoting the dissociation of N_2O into a surface oxygen atom and gas-phase N_2.

Studies have manifested that the catalytic properties of iron clusters vary with the size of metal clusters [11]. Sun et al. [12] calculated the reaction of methane activated by iron atom/dimer, and their calculation results indicated that the iron dimer Fe_2 had a stronger catalytic effect on the activation of methane than the iron atom. However, the catalytic effect of NO activation on varied size of small iron cluster is undefined. Gutsev et al. [13] et al. investigated the electronic and geometrical structures of Fe_n (n=1~6) with associative and dissociative attachments of NO, as well as the corresponding negatively and positively charged ions, as well as the corresponding negative and positive charge ions ,but the results are lack of the path of the reaction kinetics.

On the surface of iron, NO may be reduced by many iron atoms. However, as a preliminary understanding of the reaction of NO with Fe, it is necessary to know how a single NO molecular reacts with one iron atom. On this basis, the further calculation of NO molecular with iron clusters will be studied. Therefore, in present work, the reaction pathway and evolution process of NO with one iron atom will be calculated based on DFT method.

2 Computational methodology

All calculations herein were carried out by the Gaussian 09 [14] program computational package with ab initio and density functional methods. As to the DFT, Becke's

DOI: 10.1051/matecconf/20166903007

[15, 16] three-parameter hybrid exchange functional combined with the Lee–Yang–Parr [17] non-local correlation functional, namely B3LYP , was selected to perform geometry optimizations and frequency calculations in the systems. At the same time, 6-311+G(d, p) basis set was employed for the structural optimization of all reactants, intermediates, transition states, and products involved in the reactions of Fe with NO. Owing to the accuracy of calculation and reasonability of time-assuming, B3LYP has been applied in the calculations of transition metals extensively [18-20]. Kwon et al. [21] investigated the vibrational frequencies and adsorption energies of iron hydroxides on phosphates, the results show that the frequencies calculated by DFT is consistent with the frequencies of experimental IR data [22, 23].

In this paper, the B3LYP/6-311+G(d, p) basis set was utilized to calculate all geometry and energy of the stationary points of the potential energy surfaces (PES) in Fe +NO system. Meanwhile, the energy of all geometries were revised by the zero point energy (ZPE). Furthermore, in order to confirm the crucial structures of transition and intermediates, the intrinsic reaction coordinate (IRC) calculations was applied to perform the minimum reaction way.

3 Results and discussions

3.1 Fe-NO system

In the Fe+NO system, according to the previous works [11], the ground state of the iron atom is a quintet state (^5D) (the total multiplicity of Fe-NO: M = 6), the occupation of iron atom corresponding to quintet state is $3d^6 4s^2$, which is consistent with the Mebel and Hwang's work [24]. Notably, there is another occupation of the iron atom ($3d^7 4s^1$), which corresponds to the triplet state (^3F) (the total multiplicity of Fe-NO: M = 4). Tatewaki and co-workers [25] have manifested that the ^5D-^3F splitting energy is 1.32eV, which is in line with the experimental data [26] (34.1 kCal/mol).

Based on that, in the Fe+NO system, we only consider the NO activation on iron atom by quartet and sextet state. Intermediates and transition states on the quartet potential energy surface (PES) are labeled as ^{4}C and ^{4}TS. Similarly, those on the sextet state PES are labelled as ^{6}C and ^{6}TS. The optimized geometries of various intermediates and transition states for the Fe-NO system by DFT are shown in Fig. 1. While Table 1 gives the information regarding the lowest-frequency vibrational modes. The reaction pathways between the iron atom and NO are plotted in Fig. 1 and Fig. 2. Bond angle are in degrees (°), bond lengths are in angstroms (Å).

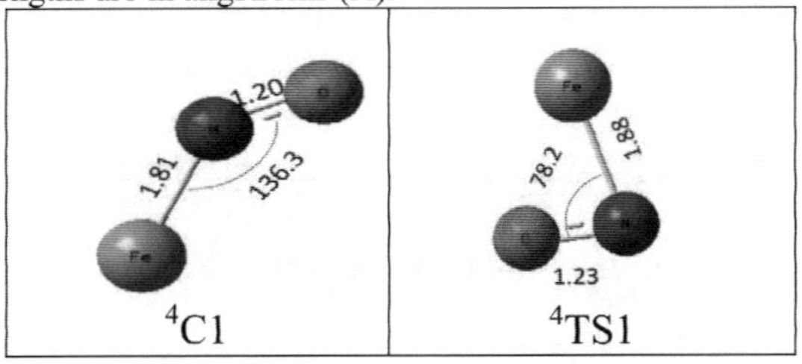

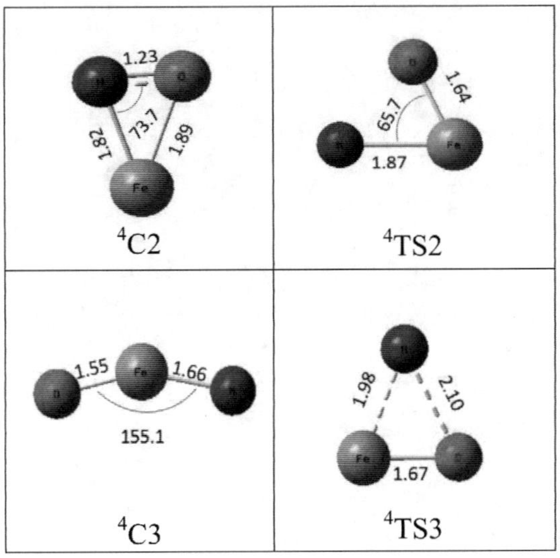

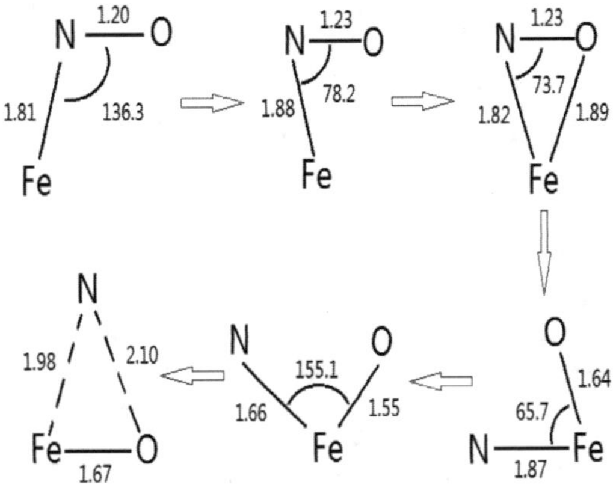

Figure 1. Optimized structures of intermediates and transition states of NO activation on iron atom on quartet state PES

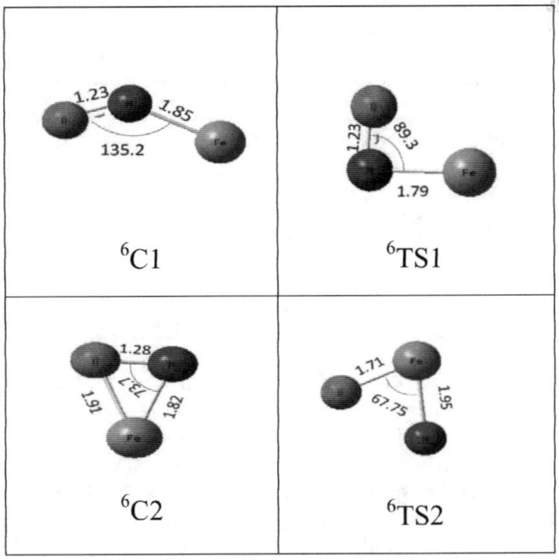

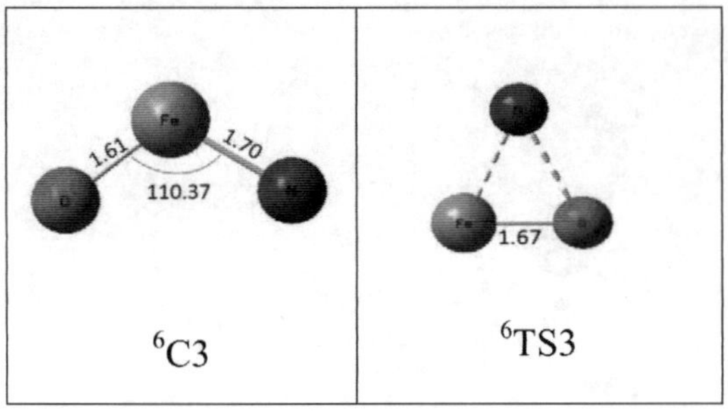

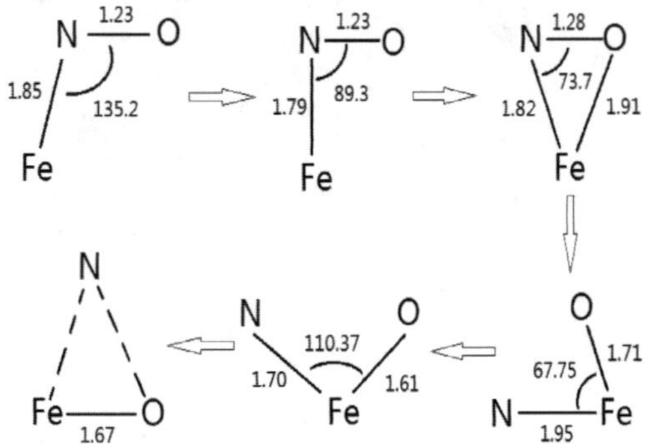

Figure 2. Optimized structures of intermediates and transition states of NO activation on iron atom on sextet state PES

3.1.1 Reaction process on two states PESs

With regard to the nitric oxide activated by the metal, previous experimental and DFT studies have demonstrated that in the presence of metals, N_2 may be the favorable product [27]. Consequently, attention was paid to the reaction pathways the state PES. Firstly, the frequencies for the saddle points on quartet/sextet states are listed in Table 1. Simultaneously, the diagram of the Fe+NO reaction on two states is shown in Fig. 3. The relative energies of various species in the reaction is show in Table 2.

Table 1. Calculated first frequencies (cm-1) of Fe+NO system at B3LYP/6-311+G (d, p) level

Quartet state		Sextet state	
Species	Freq	Species	Freq
^{4}C1	169.01	^{6}C1	269.65
^{4}TS1	696.77i	^{6}TS1	269.70i
^{4}C2	378.84	^{6}C2	431.87
^{4}TS2	522.86i	^{6}TS2	431.20i
^{4}C3	149.60	^{6}C3	212.25
^{4}TS3	433.56i	^{6}TS3	474.62i

Before desorption of the N atom from adducts, the reaction got through the cleavage of N-O. As we can learn from the Fig. 1 and Fig. 2, our calculations shows that the formation of the Fe-NO adduct was formed on quartet and sextet state channels firstly. We define the energy of the adduct on the sextet state (^{6}C1) as benchmark, then the relative energy of ^{4}C1 is -13.15 kCal/mol, which means the geometry of ^{4}C1 is more stable.

Next step on quartet and sextet PESs, ^{4}C2 and ^{6}C2 are formed via transition states ^{4}TS1 and ^{6}TS1. Correspondingly, the complexes surmount energy barrier of 31.01 kCal/mol and 38.76 kCal/mol on two PESs. While imaginary frequencies of the two transition states derive from N-O bond swing, about 696.77i cm^{-1} and 269.70i cm^{-1}, respectively. After the first transition states, the Fe-O bonds are formed. The relative energies of ^{4}C2 and ^{6}C2 are 0.67 kCal/mol and 5.96 kCal/mol.

The cleavage of the N-O bonds occur in the second transition state, which conquer energy barrier of 49.63 kCal/mol and 41.45 kCal/mol on quartet and sextet PESs, respectively. The imaginary frequencies of the second transition state (522.86i cm^{-1} for 4TS2, 431.20i cm^{-1} for ^{6}TS2) are assigned to planar rocking modes of O-Fe-N. After ^{4}TS2 and ^{6}TS2, angles of O-Fe-N get larger, about 155.1° and 110.37°. Then, third intermediates (^{4}C3 and ^{6}C3) are formed, correspondingly, the relative energies of ^{4}C3 and ^{6}C3 are 36.31 kCal/mol and 27.97 kCal/mol.

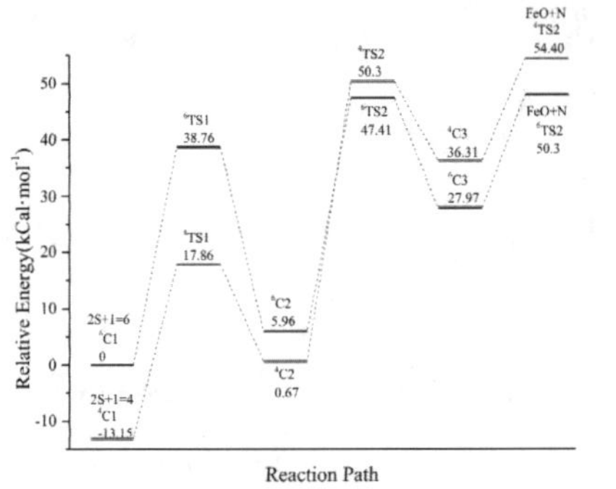

Figure 3. Diagram for the reaction of Fe+NO on quartet/sextet state PES

With the N-Fe-O angles becoming larger, the constraint on N atom from O atom is weakened. The separation of the N atom from the complexes completed in fourth transition states, The energy barriers for the cleavage of Fe-N bond are 18.09 kCal/mol and 20 kCal/mol for quartet and sextet PESs.

Table 2. Energies of various species in reaction of Fe+NO→ FeO+N on two state PESs

Species	E_r (kCal/mol)		E'(kCal/mol)	
	quartet state	sextet state	quartet state	sextet state
^{4}C1(^{6}C1)	-13.15	0	-	-
^{4}TS1(^{6}TS1)	17.86	38.76	31.01	38.76
^{4}C2(^{6}C2)	0.67	5.96	-	-
^{4}TS2(^{6}TS2)	50.30	47.41	49.63	41.45
^{4}C3(^{6}C3)	36.31	27.97	-	-
^{4}TS3(^{6}TS3)	54.40	47.97	18.09	20

3.1.2 Crossing point between two PESs

As shown in Fig. 3, there is a crossing point (CP) between the quartet and sextet state PESs. The crossing behavior in the thermal reaction of transition metal compounds was observed by Armentrout [28]. Shaik et al. [29, 30] confirmed such type of behavior and proposed that the phenomenon of two states of different multiplicities can dramatically affect the reaction mechanisms, rate constants, branching ratios, and temperature behaviors of organometallic transformations, which is termed as two-state reactivity (TSR). TSR, i.e., can make sure a thermal reaction which involves spin crossover along the reaction coordinate from reactants to products may find low energy paths for otherwise difficult processes.

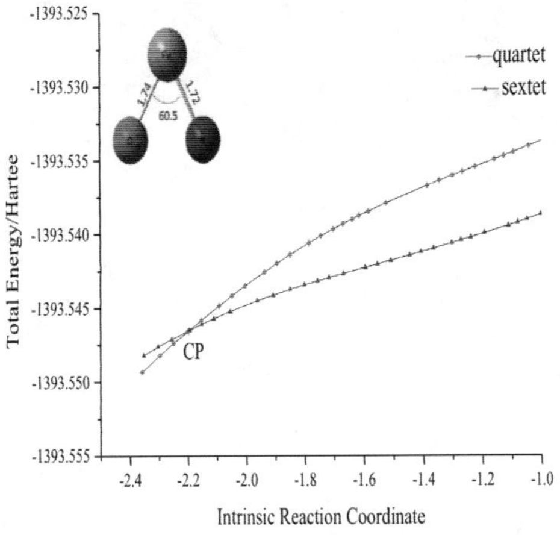

Figure 4. Potential energy cure-crossing point diagram between two spin state PESs

The crossing point (CP) and spin-crossover of TSR are important to understand the mechanism of NO activation on iron atom. Our calculations show that a crossing point exists between quartet and sextet PESs. Ulteriorly, the crossing point was verified by the intrinsic reaction coordinate (IRC) calculations. In the light of IRC calculation, the geometry of crossing point was received and listed in Fig. 4. According to the TSR, the reaction system may transform its spin multiplicity from the quartet to sextet state in the crossing region, which can lower the activation energy effectively, from 49.63 kCal/mol to 46.74 kCal/mol. Therefore, the minimum energy pathways can be described as: Fe+NO (M=4)→ 4C1→4TS1→4C2→6TS2→6C3→6TS3.

4 Conclusion

The B3LYP method combined with 6-311+G(d, p) basis set was applied to investigate the reaction of NO activation on iron atom. For the activation of NO on Fe atom, our calculations suggest that the reaction should proceed via the quartet PES, in the vicinity of CP, the reaction system will transform its spin multiplicity from the quartet to sextet state. Attribute to the CP, the activation energy can be lowered effectively, from 49.63 kCal/mol to 46.74 kCal/mol, which derives from cleavage of N-O bond. And mechanism of the reactions on quartet and sextet PESs are similar. For the activation of NO on Fe dimer, simultaneously, reaction also proceed via low spin state (octet), the reaction system may transform its spin multiplicity from the octet to dectet state in the crossing region. Also, the CP can effectively lower activation energy from 49.53 kCal/mol to 45.33 kCal/mol, deriving from breaking of Fe-O bond, mechanism of the reactions on octet and dectet PESs are similar. Results showed that the reaction of iron atom with NO was a typical two-state reaction（TSR）. In the reaction channel, the crossing point appears, which would effectively reduce the activation energy and increase the release of reaction heat, play a significant and beneficial role in the kinetic and thermodynamic aspects of this catalytic reaction.

Acknowledgments

This work was supported by National Natural Science Foundation of China (No. 51278095), which are gratefully acknowledged.

References

1. Pastor G, Dorantes-Dávila J, Bennemann K.Physica B+ C.**149** (1988).
2. Jellinek J, Garzon I L.Z. Phys. D: At., Mol. Clusters.**20** (1991).
3. Hill C L, Prossermccartha C M.Coordin. Chem. Rev. **143** (1995).
4. Alonso J A.Chem. Rev. **100** (2000).
5. Dyson P J.Coordin. Chem. Rev. **248** (2004).
6. SU Yaxin, SU Along, CHENG Hao.J China Coal Soc.**38**(s1) (2013)
7. Gradon b, Lasek J. Fuel.**89**(11) (2010)
8. SU Ya-xin, SU A-long, CHENG Hao. J. China Coal Soc. **38**(s1) (2013)
9. Blagojevic V, Orlova G, Bohme D K.J. Am. Chem. Soc. **127** (2005).
10. Heyden A, Hansen N, Bell A T, et al.J. Phys. Chem. B, **110** (2006).
11. Bansmann J, Baker S H, Binns C, et al.Surf. Sci. Rep. **56** (2005).
12. Sun Q, Li Z, Du A, et al.Fuel.**96** (2012).
13. Gutsev G, Mochena M, Johnson E, et al.J. Chem. Phys. **125** (2006).
14. Frisch, M.; Trucks, G.; Schlegel, H. B.; et al. Gaussian 09, Revision A.02; Wallingford CT: Gaussian, Inc. (2009).
15. Becke A D.Phys. Rev. A.**38** (1988).
16. Becke A, Roussel M.Phys. Rev. A.**39** (1989).
17. Lee C, Yang W, Parr R G.Phys. Rev. B.**37** (1988).
18. Ai H, Bu Y, Li P, et al.J. Chem. Phys. **123** (2005).
19. Ai H, Bu Y, Li P, et al.New J. Chem. **29** (2005).
20. Yang X-Y, Wang Y-C, Geng Z-Y, et al.Chem. Phys. Lett. **430** (2006).
21. Kwon K D, Kubicki J D.Langmuir. **20** (2004).

22. Persson P, Nilsson N, Sjöberg S.J. Colloid Interface Sci. **177** (1996).
23. Arai Y, Sparks D.J. Colloid Interface Sci. **241** (2001).
24. Mebel A M, Hwang D-Y.J. Phys. Chem. 105 (2001).
25. Tatewaki H, Tomonari M, Nakamura T.J. Chem. Phys. **88** (1988).
26. NIST atomic spectra database. <http://physics.nist.gov/asd3>.
27. Brown W A, King D A.J. Phys. Chem. B. **104** (2000).
28. Armentrout P.Science. **251** (1991).
29. Shaik S, Danovich D, Fiedler A, et al.Helv. Chim. Acta.**78** (1995).
30. Schröder D, Shaik S, Schwarz H.Accounts. Chem. Res.**33**(2000).

Investigation of the Parameters affecting CO_2 –assisted Polyaniline Polymerization

H. Noby[1,2], A. H. El-Shazly[1], M. F. Elkady[1] and M. Ohshima[3]

[1]*Chemical and Petrochemicals Engineering Department, Egypt-Japan University of Science and Technology, New Borg El-Arab City, Alexandria, Egypt.*
[2]*Material Engineering and Design, Faculty of Energy Engineering, Aswan University, Aswan, Egypt.*
[3]*Chemical Engineering Department, Kyoto University, Kyoto, Japan. 615-8510.*

Abstract. Specific Polyaniline (PANI) morphologies such as nanotubes and nanofiber are required for enhancing its performance in the various applications. CO_2 –assisted Polyaniline polymerization is a method recently used to produce these anticipated morphologies. In this study, polyaniline nanotube was prepared successfully in the presence of compressed CO_2 utilizing Aniline as a monomer and Ammonium peroxydisulfate (APS) as an oxidizing agent. The effect of both reaction temperature and the oxidizing agent feed rate on the morphology and surface area of the produced PANI was investigated. The synthesized PANI was examined by FT-IR, XRD, and BET surface area analysis. Furthermore, SEM was carried out to figure out the morphology of the prepared PANI. It was indicated that Polyaniline nanotubes PANNTs size and homogeneity were affected by the reaction temperature. The averages of the outer and inner diameters of the PANNTs at 25 °C, 45 °C, 65 °C were found to be about (120, 60 nm), (140, 65 nm), and (175, 75 nm) respectively. Also, the produced surface area was slightly augmented with the increase of the temperature. In addition, it was observed that increasing the feeding rate of the APS was associated with the reduction of the size and the surface area of the produced PANI nanotubes.

1. Introduction

The porosity and morphology of Polyaniline (PANI) are the most effective properties that concern its performance [1, 2]. PANI can be easily synthesized in different morphologies as nanofiber, nanoparticles, nanorods, and nanotubes. Besides, Insignificant PANI surface area may restrict the use of PANI powder in many applications, but it may be overcome by manufacturing PANI nanotubes (PANNTs) [3]. Therefore, lately a great attention has been focused on producing PANNTs with and without the aid of templates [1]. Furthermore, many dissimilar techniques, such as template synthesis, interfacial polymerization [4], and self- assembly have been publicized to produce PANNTs [5]. In order to prepare the PANI in advanced morphologies and in a high porous structure, polymerization assisted with subcritical or supercritical CO_2 was used [6-9]. Polymerization of PANI in subcritical and supercritical CO_2 still needs more investigations to demonstrate the influence of different parameters on the produced morphology. In this experimental investigation, PANNTs was prepared by CO2-assisted polymerization. Also, the influence of the reaction temperature and the APS feed rate into the reaction vessel on the attained PANI Morphology and Surface area were acquired.

2. Experimental

2.1 Materials

Aniline, Ammonium peroxydisulfate (APS), Acetic acid, Hydrochloric acid, Ethanol, and Methanol (Wako Pure Chemical Industries, Japan) were used without further purification. Freshly distilled water was used for the preparation of all aqueous solutions and the washing process.

2.2 Polyaniline preparation

The PANNTs were prepared according to the process reported previously [8]. Aniline (0.2 M) solution in 25 ml of distilled water containing 3 ml of acetic acid and 0.25 M of APS solution in another 25 ml of distilled water were prepared. As shown in Fig. 1, the aniline solution was placed in an autoclave (high pressure vessel) connected to a CO_2 pump. Firstly, the autoclave located in an oven to control the reaction temperature. The

 DOI: 10.1051/matecconf/20166904001

temperature varied as 25, 45, 65 °C to study its effect on the prepared PANI. The CO_2 pressurized into the reactor causing a pressure rise of 9 MPa for 1 hrs. Secondly, APS solution was fed into the vessel by another pump with a constant feeding changed as 0.5, 1, and 1.5 ml/min. Then, the mixture was left to polymerize for 2 hrs. After 2 hrs of polymerization, a valve was opened to release the CO_2 from the main vessel. The prepared polymer was collected, filtered by centrifugation at 5000 rpm for 5 min, and washed by 1 M HCl, Ethanol solutions, and distilled water. Then it was dried at 80 °C for 24 hrs.

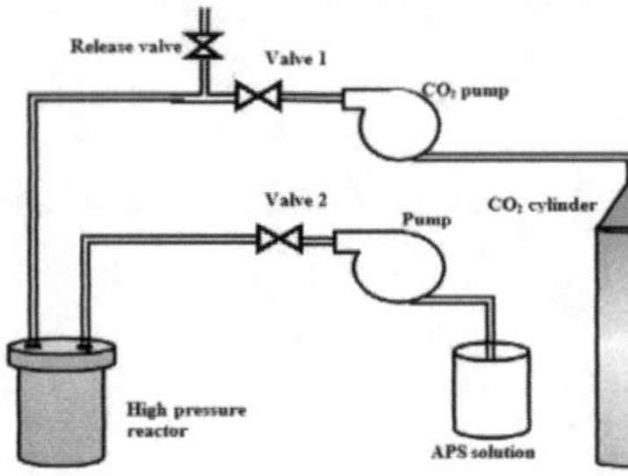

Figure 1. Schematic diagram for the methodology used in PANI CO_2-assisted polymerization.

2.3 Characterization of the prepared PANI

IR spectrometer (Vertex 70, Bruker scientific instruments, Germany) with a wavelength range from 4000 to 400 cm^{-1} was employed to show the FT-IR of the synthesized PANI. The spectrum in the LabX XRD-6100 Shimadzu, Japan, X-ray diffraction was used to show the crystallinity of PANI. N_2 adsorption/desorption isotherms at 77 K and the Brunauer-Emmett Teller (BET) surface area analysis were conducted using (BelsorbminiII, BEl Japan Inc., Japan). All the samples were degassed under vacuum at 150 °C for 2 hrs. Barrett, Joyner and Halenda (BJH) method was employed to determine the pore size distribution and total pore volume. The morphology was observed by scanning electron microscopy SEM (JEOL JSM-6010LV, Japan).

3. Results and discussion

Regarding the improvement of the PANI surface area and attaining advanced PANI morphology, the polymerization temperature and the feed rate of the ABS impact on the manufactured PANI was observed by SEM, BET surface area analysis, XRD, and FT-IR.

3.1 Morphological structure

Fig. 2 shows SEM images of the PANNTs prepared at different temperatures and APS feed rates, where the nanoscale tube-like structure, i.e., Polyaniline Nanotubes (PANNTs) was observed. The effect of temperature on the morphological structure at constant APS flow rate of 0.5 ml/min was determined in Fig. 2 (a, b, and c). Uniform nanotubes of PANI were noticed at low reaction temperature. It was noticed that the uniformity of the nanotubes diminished with the surge of the temperature. Therefore, some nanoparticles were observed at 45 and 65 °C SEM images.

As seen, a rise of the tubular structure diameter with the increase of temperature was noticed. According to the SEM images, the averages of the outer and inner diameters of the PANNRs at 25 °C, 45 °C, 65 °C were found to be about (120, 60 nm), (140, 65 nm), and (175, 75 nm) respectively. The length of the produced structure was in the range of (1- 2.5 μm). PANI has great thermal stability, so as expected the increasing of the temperature reduces this thermal stability which facilitate the diffusion of the CO_2 molecules into the entire region of PANI before it released. Besides the thermal stability, rising the temperature to 45 or 65 °C will change the status of the CO_2 from sub-critical to supercritical Sc-CO_2. Sc-CO_2 has better diffusivity into the different solutions compared with the sub-critical; this will resulted in high surface area. Therefore, the BET surface area analysis confirmed this phenomena and showed better surface area with augmentation of the temperature.

Figure 2. SEM images of the prepared PANNTs with rate of 0.5 ml/min at (a) 25 °C, (b) 45 °C, and (c) 65 °C.

The effect of APS feed rate on the morphological structure was observed in Fig. 2a and 3 (a and b). Almost, uniform nanotubes of PANI were noticed at all APS feed rates. As shown, the rise of the APS feed rate was accompanied by more homogeneous nanotubular structure and diminution of the diameter size. The average outer and inner diameters of the PANNRs at 0.5, 1.5, and 1.5 ml/min were found to be about (160, 75 nm), (140, 65 nm), and (135, 60 nm) respectively. The enlarging of the APS flow rate may increase the contact points between the monomer and the oxidizer which creates more PANI nucleation points. This may cause the lessening of the diameters size.

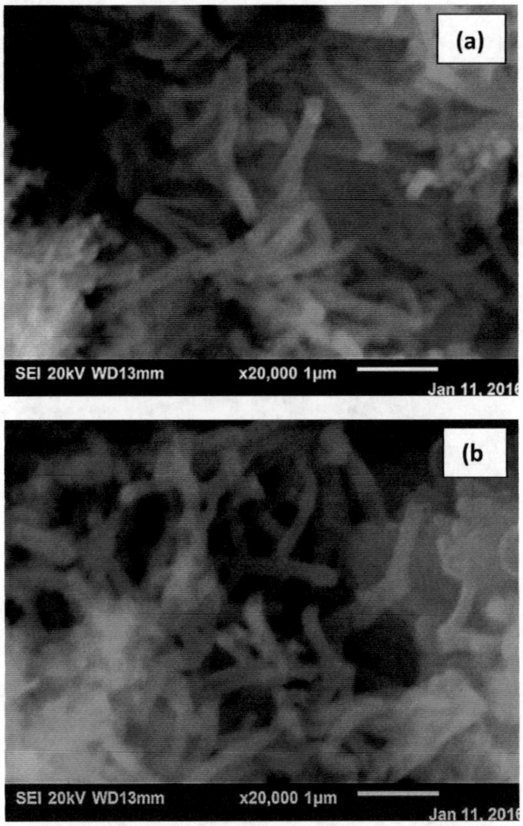

Figure 3. SEM images of the PANNTs prepared at uniform room temperature and (a) 1 ml/min, and (b) 1.5 ml/min.

3.2 Surface area and Pore size

Table 1 shows the BET surface area, the total pore volume and the average pore size of the produced PANNTs different samples. N_2 adsorption/desorption isotherm and pore size distribution of the produced PANI at different reaction temperature were shown in Fig. 4 (a, b, and c). The prepared PANI structures showed a hysteresis loop H3 type according to the IUPAC classification [10]. Also, a wide range pore size distribution (from 1.4 to 70 nm) was observed for all the tested temperatures, which can be considered a macroporous structures according to the IUPAC pores classification. Based on the data of Table 1, the BET surface area of the produced PANNTs samples were 22.13, 25.71, and 28.9 (m^2/g) at 25 ^oC, 45 ^oC, 65 ^oC respectively.

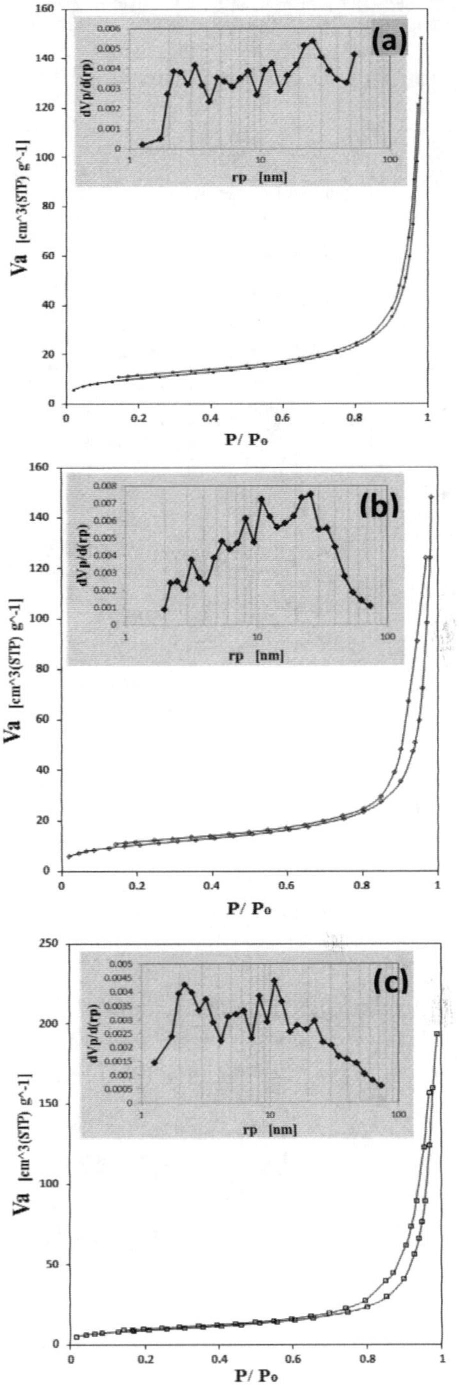

Figure 4. N_2 adsorption-desorption isotherm and pore size distribution of PANNTs with APS feed rate of 0.5 ml/min at (a) 25 ^oC, (b) 45 ^oC, and (c) 65 ^oC.

Fig. 4a and 5 (a, and b) represent the N_2 adsorption/desorption isotherm and pore size distribution of the synthesized PANNTs at various APS feed rates at a uniform reaction temperature of 25 °C. The hysteresis loop changed from H3 type to H4 with the rise of the feeding rate. This may related to the nanoporous pores attained at the material as confirmed by Table 1. Moreover, the BET surface area of the produced PANNTs samples were diminished until it reached 8.47 (m^2/g) at APS feeding rate of 1.5 ml/min.

Table 1. Surface area, total pore volume, and average pore diameter of different PANI samples.

Parameters	Reaction temperature °C			APS feed rate ml/min		
	25	45	65	0.5	1	1.5
Surface Area (m^2/g)	22.13	25.7 1	28. 9	22.1 3	18.5 6	8.47
Total Pore Volume (BJH) (cm^3/g)	0.229	0.28 6	0.2 9	0.22 9	0.18 7	0.082
Average Pore Size (BJH) (nm)	25.71	27.6 3	23. 6	25.7 1	24.7 6	2.5

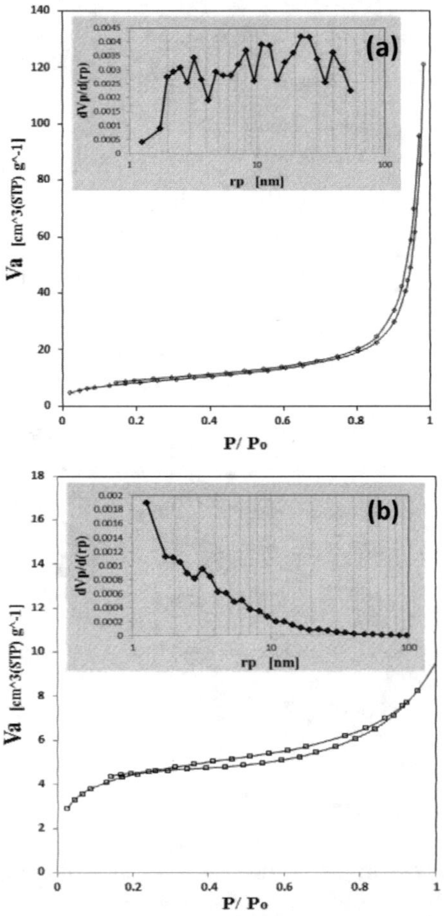

Figure 5. N_2 adsorption-desorption isotherm and pore size distribution of PANNTs at room temperature and (a) 1 ml/min, and (b) 1.5 ml/min.

3.3 Crystalline structure

The crystallinity of the produced PANNTs with different temperatures and feed rates were analyzed by XRD. PANNTs show a characteristic emeraldine salt form. Fig. 6a shows XRD patterns of PANNTs prepared at different temperatures. The XRD pattern of PANI has been reported by several researchers [11-13]. PANI is a semi-crystalline polymer, which has both amorphous and crystals phases. The XRD pattern shows two peaks: one was centered on 2θ=19.38° which attributed to a periodicity parallel to the produced PANI chain and the other was at 2θ=25.7° which associated to a periodicity perpendicular to the polymer chain [12, 13].

XRD patterns of PANNTs prepared at different APS feed rate were figured out in Fig. 6b. Although, no changes were clearly noticed between the XRD peaks of the all attained PANI structure, the crystallinity slightly increased with the rise of the APS feed rate. The high tapered peaks attained with the high feed rate PANI may shows this crystallinity rise.

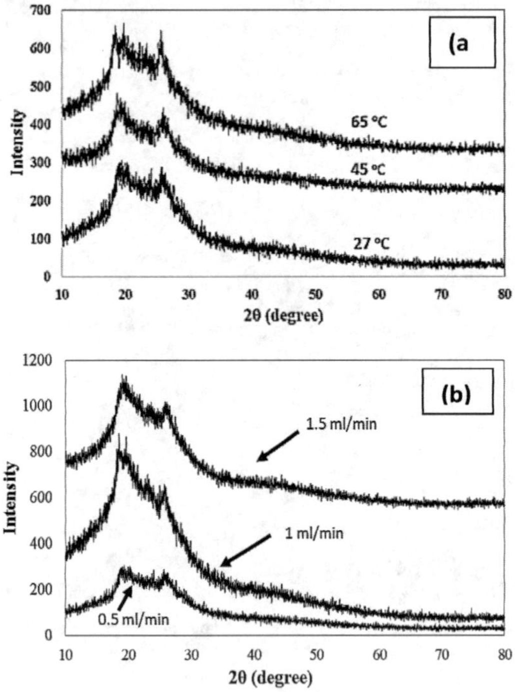

Figure 6. XRD of the prepared PANNTs at (a) different temperatures and uniform feed rate of 0.5 ml/min, (b) different APS feed rates and uniform temperature of 25 °C.

3.4 Chemical structure

The FT-IR of the manufactured PANNTs at various temperatures and APS feed rates were shown in Fig. 7a and 7b. The FT-IR spectra obtained in this study shows the conformity with the previously reported outcomes [11-16]. For all the PANNTs prepared, the characteristic Peaks of the benzenoid and quinoid units performed around 1514 and 1595 cm^{-1} which confirm the existence of the emeraldine salt form due to the approximately

DOI: 10.1051/matecconf/20166904001

equal benzenoid and quinoid peaks intensity. The peaks at 719 and 1310 cm^{-1} are ascribed to aromatic C—H out-of-plane bending and C—N stretching of the secondary aromatic amine vibrations respectively. The band at 837 cm^{-1} is associated with C—C and C—H for the benzenoid unit. The broad bands in the range of 3000–3500 cm^{-1} are attributed to the free (non-hydrogen bonded) N—H stretching vibration.

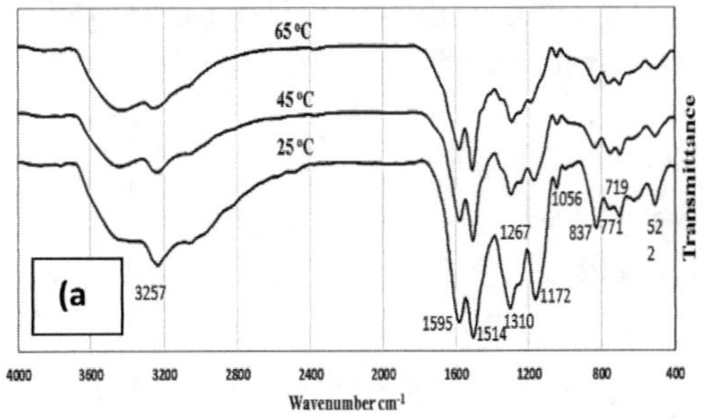

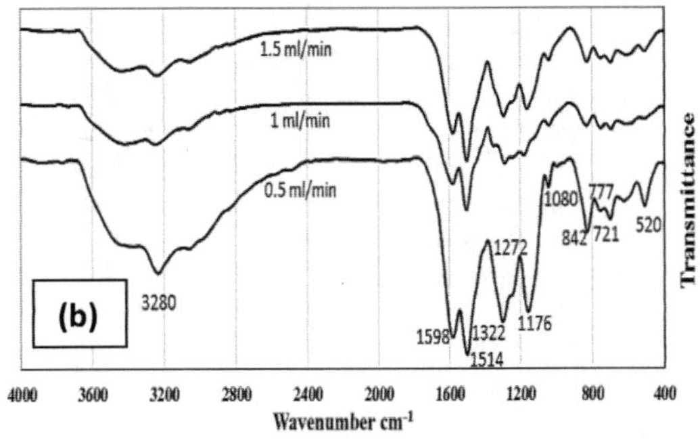

Figure 7. XRD of the all prepared PANNTs at (a) different temperatures at APS feed rate 0.5 ml/min, (b) different APS feed rates at temperature 25 °C.

4. Conclusions

CO_2-assisted polymerization technique was successfully employed to manufacture PANI nanotubes. The main reactants were the Aniline monomer and APS as an oxidizing agent. At first, the compressed CO_2 was diffused into the Aniline solution, and then the APS solution was pumped with a uniform flow rate. The effect of the APS feeding rate and the reaction temperature on the produced PANI morphology and surface area was acquired. According to the SEM, PANNTs size and homogeneity were affected by the reaction temperature variation. The averages of the outer and inner diameters of the PANNRs were assessed to be around (120, 60 nm), (140, 65 nm), and (175, 75 nm) at 25 °C, 45 °C, 65 °C respectively. Moreover, the measured surface area was slightly enlarged with the rise of the temperature. In addition, it was observed that growing the feeding rate of the APS was associated with the reduction of the surface area and size of the synthesized PANI nanotubes.

Acknowledgment

We would like to acknowledge Egypt Japan University for support. We would also like to thank and acknowledge all members of *Materials Process Engineering lab*, *Kyoto University* for their help to conduct the experiment.

References

1. X. Han, H. Qiu, F. Qiu, and J. Yang, Appl. Surf. Sci, (2016).
2. M. Nandi, R. Gangopadhyay, and A. Bhaumik, Microporous Mesoporous Mater., **109**, 239–247, (2008).
3. C. Xu, H. Chen, and F. Jiang, Colloids Surfaces A Physicochem. Eng. Asp., **479**, 60–67, (2015).
4. C. Oueiny, S. Berlioz, and F. X. Perrin, Chem. Phys., **18,** 3504–3509, (2016).
5. Z. Huang, E. Liu, H. Shen, X. Xiang, Y. Tian, C. Xiao, and Z. Mao, Mater. Lett., **65**, 2011, 2018, (2015).
6. Q. Pham, J. Kim, S. Kim, Synth. Met., **160,** 394–399, (2010).
7. J. Du, J. Zhang, B. Han, Z. Liu, and M. Wan, Synth. Met., vol. **155**, 523–526, (2005).
8. H. Noby, A.H. El-Shazly, M.F. Elkady, M. Ohshima, Int. Conf. Chem., Biological and Env. Eng. (ICBEE 2015), (Milan, Italy), 14–15 Sep. (2015).
9. H. Noby, A.H. El-Shazly, M.F. Elkady, M. Ohshima, 13th Int. Conf. Tutorials Adv. Foam Mat. Tech. (FOAMS® 2015), Kyoto, Japan, 10–11 Sep. (2015).
10. K. Kaneko, J. Membrane Sci. **96,** 59-89, (1994).
11. Najim, T.S., Salim, A.J., Arabian J. Chem., (2014), http://dx.doi.org/10.1016/j.arabjc.2014.02.008.
12. L. Ai, J. Jiang, R. Zhang, Syn. Met. **160**, 762–767, (2010).
13. L. Ren, G. Zhang, J. Wang, L. Kang, Z. Lei, Zhongwen L., Z. Liu, Z. Hao, Z. Liu, Electro. Acta **145,** 99–108, (2014).
14. M. Ayad, G. El-Hefnawy, S. Zaghlol, Chem. Eng. J. **217,** 460–465, (2013).
15. J. Wang, K. Zhang, L. Zhao, Chem. Eng. J., **239**, 123–131, (2014).
16. F. Yang, M. Xu, Sh. Bao, H. Wei, H. Chai, Electro. Acta., **137** (2014) 381–387.

MATEC Web of Conferences **69**, 04002 (2016) DOI: 10.1051/matecconf/20166904002
ICCPE 2016

Microencapsulation of citronella oil by complex coacervation using chitosan-gelatin (b) system: operating design, preparation and characterization

Fitrah Rabani Abdul Aziz, Junaidah Jai, Rafeqah Raslan and Istikamah Subuki

Faculty of Chemical Engineering, Universiti Teknologi MARA, 40450 Shah Alam, Selangor, Malaysia

Abstract. Citronella oil (CO) can be an effective mosquito repellent, but due to its nature which having high volatility, oils rapidly evaporates causing loss of efficacy and shorten the repellent effect. Therefore, microencapsulation technology was implemented to ensure the encapsulated material being protected from immediate contact with environment and offers controlled release. In this study, microencapsulation of CO was done by employing complex coacervation using chitosan-gelatin (B) system and utilized proanthocyanidins as the crosslinker. Remarkably, nearly all material involved in this study are from natural sources which are safe to human and environment. In designing operating process condition for CO encapsulation process, we found that wall ratio of 1:35 and pH 5 was the best operating condition based on zeta potential and turbidity analysis. FT-IR analysis found that gelatin-B had coated the CO droplet during emulsification stage, chitosan started to interact with gelatin-B to form a polyelectrolyte complex in adjust pH stage, CO capsules solidified at cooling process and were hardened during crosslinking process. Final product of CO capsules after settling process was identified at the top layer. Surface morphology of CO capsules obtained in this study were described having diameter varies from 81.63 μm to 156.74 μm with almost spherical in shape.

1 Introduction

Citronella oil (CO) is an essential oils extracted from aromatic grasses of Citronella grass. It is classified in trade into two types; Ceylon CO, obtained from *Cymbopogon nardus* (inferior type), while Java CO obtained from *Cymbopogon winterianus* (superior type). This plant can be found growing wild in most tropical Asian countries including Malaysia. CO is comparatively high demand due to its usage in traditional medication. It also widely used in perfume industry, soap manufacturing, cosmetics, flavouring industry and health purposes [1]. In addition, CO also shows good potential usage in food packaging and textile industry [2, 3]. In fact, Environmental Protection Agency (EPA) itself has classified CO as a biopesticide with a non-toxic mode of action [4, 5]. Therefore, CO can be found in dozens of registered pesticide products such as sprays, lotions, candles, sunscreen, wristbands, and pet's collars [4, 6]. This makes essential oil (EO) from Cympobogon species such CO is popular in insect or mosquito repellent formulations. Candles and incense containing CO are sold as insect repellents in some countries like the United States [7]. There was also a study done by [8] which utilized CO in formulation of mosquito repellent water-based paint. It was suggested by [9] and [10] that CO demonstrates good efficacy against mosquitoes in concentrations ranging from 0.05% to 15% (w/v) alone or in combination with other natural or commercial insect repellent products. Although synthetic chemical *N,N*-diethyl-3-methylbenzamide (DEET) is currently the most effective mosquito repellent, but due to adverse effects reported such include skin eruption in children and irreversible damage to the ecosystem from nondegradable compound in DEET, development of effective non-DEET alternatives is needed [7]. Thus, citronella is one of the most widely used natural repellents on the market nowadays and was used at concentrations of 5-10%. It was stated in [11] that most of the EO-based repellents tend to give short lasting protection for less than 2 hours. However, CO was found to give 2 hours repellency and be the most effective mosquito repellent compared to the other thirty eight EOs that have been studied in [11]. Initially, CO can be as effective as DEET in repels mosquito or insect, but due to its nature, oils rapidly evaporate causing loss of efficacy and leaving the user unprotected [12].

Microencapsulation technology is well known as a process of packaging micronized liquid or solid material in the form of microcapsules. This process give the encapsulated material affirm protection from immediate contact with environment besides provide controlled release [13]. Numerous manufacturing approaches have been adopted in producing microcapsules and the most common used processes include complex coacervation, spray drying, centrifugal extrusion, air suspension coating, pan coating, emulsion hardening process, polymer-polymer incompatibility, interfacial polymerization and in-situ polymerization [14, 15]. By comparing with various microencapsulation techniques, [16] claimed that coacervation is one of the simplest and cheapest ways to form microcapsules. It is a promising technique due to high payloads achievable (up to 99%) [17]. Complex coacervation simply can be described as method where two or more molecules type of opposite charge, are joined as the walls and then disperse the core substance into the aqueous solution of wall materials. When the conditions are right, the wall materials will attract with each other because of the opposite charges, and then embed the core material. It is widely used to prepare microcapsules for controllable drug release because of its simplicity and low pollution [18].

Systems containing both cationic and anionic hydrophilic colloids can induce coacervation. Thus, complex coacervation is expected to occur between chitosan, a water soluble cationic polysaccharide which is soluble in dilute acids at pH 5-6, and type-B gelatin, a protein which is negatively charged at pH values above its isoelectric point which is around pH 4.7–5.2 [19]. Work done by [20] claimed that water stability of chitosan-gelatin (B) blends was found to decrease with greater gelatin content, thus chemical cross-linking is necessary. Cross-linking is a method used to branch chemical and physical junctions within polymeric materials [21]. The most common cross-linker used associates to gelatin are formaldehyde and glutaraldehyde [21, 22]. Other material that had also been used as cross-linker in chitosan-gelatin system are polyethylene oxide (PEO), polyvinyl acetates (PVA) and 1-ethyl-3-(3-dimethylaminopropyl)-carbodiimide (EDC) [22]. However, this kind of synthetic cross-linker unfortunately can cause physiological toxicity [23]. Therefore, natural and nontoxic cross-linker such as Proanthocyanidins (PA) is preferable in order to strengthen and stabilize chitosan-gelatin blend. This material which majorly contain in fruits and vegetables is about 120 times less toxic than glutaraldehyde [22].

The objectives of this paper can be divided into four: (1) to design operating condition for complex coacervation of CO using chitosan-gelatin(B) system; (2) to encapsulate CO using this method; (3) to investigate any changes in chemical structure of material throughout each stage of production process; (4) to analyze surface morphology of prepared CO capsules.

2 Methodology

2.1 Material

Citronella oil (Java, pure essential oil) was purchased from BF1 (Malaysia). Gelatin-B (type B, 260 bloom, from bovine) was supplied from Halagel Sdn. Bhd. (Malaysia). Chitosan (70-97% deacetylation, 10-1500 mPa.s) from R&M Chemicals (Malaysia) was used in this experiment. Oligoproanthocyanidin from grape seed extract (GSE-OPC) contains approximately 80% Proanthocyanidin was purchased from Kordel's (Malaysia). Other chemicals in analytical grade such as glacial acetic acid (CH_3COOH), sodium hydroxide (NaOH) and de-ionized water (DIW) were used throughout this experiment.

2.2 Designing wall ratio and operating pH

Both wall material solutions were prepared according to [24], where solutions of 1% (w/v) gelatin-B (Gel-B) solution were prepared by 30 minutes soaking the protein in DIW prior to 1 hour stirring at 50°C until fully dissolved. Whereas 1% (w/v) chitosan (Chi) solutions were prepared by dissolving analogous amount of Chi in 1% (v/v) CH_3COOH aqueous solution for more than 12 hours at room temperature. Zetasizer® Nano Series (Malvern Instruments Ltd., Worcestershire, UK) was used to measure zeta potential value of both polymeric solutions at pH range from 4.0 to 5.6 with 0.2 increments. In order to determine operating pH and wall ratio, turbidity analysis of wall mixture were conducted. Series of wall ratio was developed from zeta potential data. In a small amount, the wall material solutions were mixed according to the developed ratio and been stirred at 50°C (500 rpm) for 30 minutes. The mixed sample solution was then analyzed using spectrophotometers (UviLine 9400, Secomam, France) at 600 nm wavelength as according to [25].

2.3 CO encapsulation process

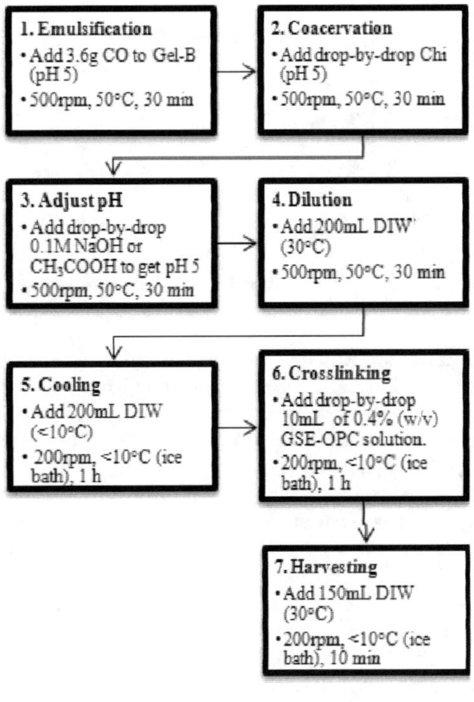

Figure 1. Citronella oil (CO) encapsulation process.

Operating condition for CO encapsulation process by complex coacervation using Chitosan-Gelatin(B) system was shown in Fig. 1. This study employed 350 mL Gel-B and 10 mL Chi respectively. The process steps were established by taking [16, 22], [26]-[28] as references. Few drop of mixture from each process step was analyzed by using optical microscope (RZ-5, Meiji Techno, Japan) and Fourier Transform Infra-Red, FT-IR (Perkin Elmer / Spectrum One). The microscopic images were taken at 30x magnification whereas FT-IR spectra were recorded by absorbance (A) mode in wavelength range 4000 – 515 cm^{-1}.

2.4 CO capsules product collection and analysis

After harvesting process, the solution was transferred into a separating funnel and undergoes settling process for more than 12 hours. Three layers were obtained which top layer (contained CO capsules in super dilute excess polymer mixture), middle layer (contained dilute excess polymer mixture) and bottom layer (contained concentrate excess polymer mixture). Aliquots for each layer were taken for analysis using optical microscope and FT-IR. The bottom layer and middle layer were drained off and the top layer was rinsed with distilled water 3 times in order to get final product of CO capsules.

3 Result and discussion

3.1 Zeta potential and turbidity analysis

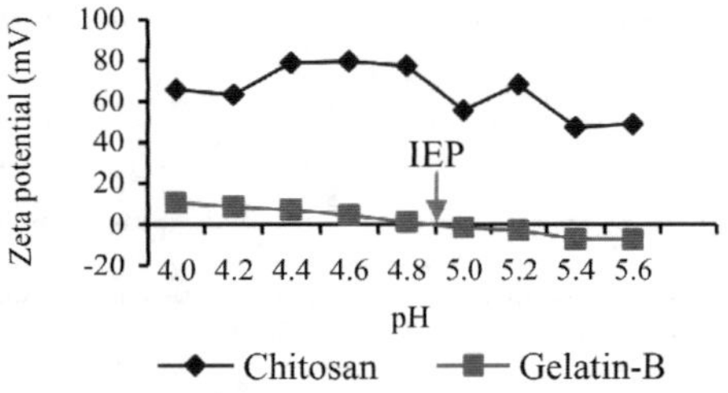

Figure 2. Zeta potential values of 1% (w/v) Chitosan and Gelatin-B respected to pH.

In designing operating condition for complex coacervation process, it is essential to know the charge for both of the wall materials. Thus, Fig. 2 shows zeta potential values of 1% (w/v) Chi and Gel-B respected to pH ($p<0.05$). Based on this data, we observed that all zeta potential values for Chi are in positive value which indicates that chitosan is positively charge and acts as polycation within range of pH 4.0 to 5.6. This result aligned with [29] which claimed that chitosan being a cationic polysaccharide in neutral or basic pH conditions but insoluble in aqueous media. However, in acidic condition, chitosan is soluble in water due to protonation of amino groups. Meanwhile, zeta potential values for Gel-B shows both positive and negative value respected to pH. This result signified that gelatin-B having amphoteric properties which its solution can be either positive or negative charge depending on pH. This statement was supported in [30] which claimed that gelatin in general is a polyampholyte obtained from denatured collagen. In this study, the isoelectric point (IEP) which indicates the neutral charge [23] for gelatin was figured out at approximately pH 4.9. As according to [31], in order for complex coacervation takes place to produce capsule, oppositely charged polyelectrolyte are present and mixed in aqueous media causing spontaneous liquid-liquid phase separation. Therefore, the designed operating pH was restricted between pH 5.0 to 5.6 so that both Chi and Gel-B are able to present as polycation and polyanion respectively.

Table 1. Wall ratio value from ratio of Chitosan (Chi) and Gelatin-B (Gel-B) approximate equal charge

pH	Chi (mV)	Gel-B (mV)	Gel-B approximate equal charge (mV)	Multiplier	Chi:Gel -B
5.0	55.53	-1.60	-55.52	34.70≈35	1:35
5.2	68.23	-3.00	-68.22	22.74≈23	1:23
5.4	47.23	-7.22	-47.22	6.54≈7	1:7
5.6	49.00	-7.33	-48.96	6.68≈7	1:7

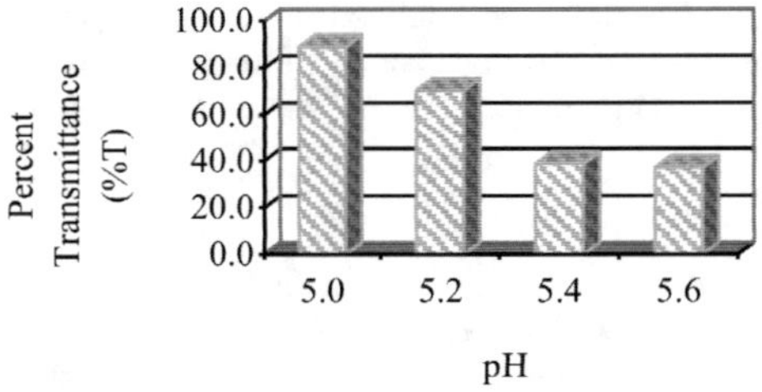

Figure 3. Percent transmittance of Chitosan-Gelatin (B) mixture against pH.

Series of wall ratio for pH 5.0, 5.2, 5.4 and 5.6 were developed by considering on zeta potential value of both Chi and Gel-B. By using multiplier, approximate equal charge of Gel-B were obtained and wall ratio from pH 5.0 to pH 5.6 are presented in Table 1. The turbidity analysis provides the final operating pH and wall ratio for complex coacervation of CO using Chitosan-Gelatin (B) system. Similar finding with [25] work was obtained where initially both Chi and Gel-B solutions appeared almost transparent in colour when they are alone as solutions. The transparency decreased when these two polymers were mixed together forming turbid or cloudy mixture as pH increased. In this case, turbidity represents the formation of insoluble complexes and causing increase in absorbance [24]. Since transmittance having opposite relationship with concentration and absorbance, thus when transmittance is low, the turbidity will be high [32]. Fig. 3 represents the percent transmittance data for Chitosan-Gelatin (B) mixture against pH. Based on this figure, it shows that as pH increased, the percent transmittance of Chitosan-Gelatin (B) mixture was decreased resulting increment in turbidity. According to [33], charges on the polyelectrolytes must be sufficiently large to cause significant electrostatic interactions for complex coacervation to occur, but too large charges can cause precipitation. In selecting best wall ratio and operating pH, condition where Chitosan-Gelatin (B) mixture become precipitate need to be avoided. In this study, we observed that Chitosan-Gelatin (B) mixture at pH 5.4 and 5.6 showed remarkable precipitation. Therefore, only pH 5.0 and pH 5.2 left for the selection of final operating pH and wall ratio. Based on the turbidity analysis done by [25], the operating wall ratio was selected by taking the maximum turbidity or highest percent transmittance of chitosan and gelatin-B mixture which was at ratio 1:15. The maximum percent transmittance was observed at pH 5.0 when the ratio of chitosan to gelatin was 1:35. Therefore, the operating process condition was set at wall ratio of 1:35 and pH 5. Comparable operating condition was applied by [19], which emulsified the oil phase on the mixture of polymers (gelatin-B and chitosan) at pH 5.0.

3.2 FT-IR spectra

Analysis on the changes in chemical structures of each CO capsules was done by using FT-IR. The FT-IR spectra obtained at each process stage of CO encapsulation process are represented in Fig. 4. FT-IR spectra at emulsification stage is shown in Fig. 4(a) represent for CO, Gel-B at pH 5 and emulsion of CO+Gel-B. CO showed strong sharp peak at 2923.03 cm^{-1} and 2853.69 cm^{-1} for C-H stretching of alkanes and methyl C-H bending of alkanes at 1455.88 cm^{-1} and 1374.54 cm^{-1}. Peak at 1229.92 cm^{-1} and 1108.75 cm^{-1} represent C-O stretch of alcohol and few peak observed at wavelength around 888.02 cm^{-1}, 824.53 cm^{-1} and 721.35 cm^{-1} for alkenyl C-H bending. Since this study used Java CO from *Cymbopogon winterianus*, which is rich in monoterpene alcohol such as citronellal, geraniol and citronellol [10], therefore alchohol and alkanes peak appearance in FT-IR spectra proven the characteristic of CO. Gel-B spectra showed strong broad peak at 3309.73 cm^{-1} which represent normal polymeric O-H stretching and N-H stretching of amines and amides. Besides, Gel-B spectra also showed a broad weak peak at 1737.90 cm^{-1} represent C-O stretching of ester and a broad medium peak at 1635.37 cm^{-1} indicating N-H bending of amines and C-O stretching of amides respectively. Methyl C-H bends appeared at peak 1365.92 cm^{-1} and a broad but weak peak at 1267.03 cm^{-1} for C-O stretching of ester group also indicating Gel-B characteristic. Spectra for emulsion of CO+Gel-B showed disappearance of all CO peaks left alone peak at 3308.51 cm^{-1}, 2108.80 cm^{-1} and 1635.40 cm^{-1} which refer to Gel-B peak without the presence of peak at 1737.90 cm^{-1}, 1365.92 cm^{-1} and 1217.03 cm^{-1}. The disappearance of all CO peak and some of Gel-B peak explained that Gel-B had acted as polyanions from C-O stretching of carbonyl group, and then coated the CO droplet during emulsification stage.

Coacervation stage FT-IR spectra was shown in Fig. 4(b) for spectra of emulsion of CO+Gel-B, Chi at pH 5 and spectra after addition of Chi. For Chi at pH 5, the spectra showed certain similarity to Gel-B spectra as mentioned by [23], where a broad strong peak appeared at peak 3308.80 cm^{-1} showed normal polymeric O-H stretching and N-H stretching of amines and amides. A broad weak peak at 1737.53 cm^{-1} represent C-O stretching of carbonyl group and a broad medium peak at 1636.09 cm^{-1} indicating N-H bending of amines and C-O stretching of amides in Chi structure. Peak at 1365.82 cm^{-1} and 1217.00 cm^{-1} for C-O stretching showed presence of cyclic ether group in Chi. FT-IR spectra for CO+Gel-B+Chi showed similar pattern with FT-IR spectra for CO+Gel-B after addition of Chi. This explained that no significance chemical changes happened at coacervation stage. However in adjust pH stage shown in Fig. 4(c), two sharp medium peak at 1365.89 cm^{-1} and 1217.01 cm^{-1} of cyclic ether C-O stretch from Chi peak was noticeable showing the polyanionic-polycationic interaction between Gel-B and Chi to form polyelectrolyte complex of Chi-Gel (B). Moreover, there are weak peak noticed around carbonyl region at 1800 cm-1 to 1700 cm-1 which proven this interaction as according to [23] explanation. Finding from [22] FT-IR analysis (using transmittance mode) described that C=O group of gelatin interacted with N-H groups of chitosan resulting intensity increased to N-H bending and decreased to C-O stretching vibration. In this study, peak at 1635.57 cm^{-1} of N-H bending showed decreased intensity of absorbance whereas at peak 1217.01 cm^{-1} of C-O stretching showed increased intensity of absorbance. It seem that opposite result was obtained if compared to previous description, however similar finding was actually obtained since absorbance and transmittance unit for FT-IR analysis having opposite relationship.

In dilution stage represent in Fig. 4(d), the FT-IR spectra obtained showed indifference on peak observed from previous stage (adjust pH), since addition of water only reducing the concentration of the mixture hence reducing the intensity of the spectra obtained without changing the chemical structure. It was explained in [34] that adding sufficient water is for the purpose of maintaining fluidity and ease of agitation after the addition polyanion or polycation (coacervation stage). There was a peak notable at 1737.59 cm^{-1} in cooling stage when comparing with previous stage of dilution. Based on Fig. 4(e), the peak was corresponding to C-O stretch of ester group from Gel-B. It was due to gelation of Gel-B upon cooling at temperature 10°C. As according to [35], gelatin solution with a concentration more than 1% will increase its viscosity drastically when cooled to 35-40°C, and form a gel when cooled further. Jason and Kalota (1996) also supported that complex coacervation emulsion mixture need to be cooled to a temperature in the range of from 5°C to about 15°C in order for the gelatin to gel or solidify into microcapsules. Fig. 4(f) showed FT-IR spectra at crosslinking stage for mixture at cooling stage, GSE-OPC alone, and after the addition of crosslinking agent, GSE-OPC. Based on the figure, GSE-OPC showed a broad strong peak at 3309.04 cm^{-1} for hydrogen bonded O-H stretch of phenol and a broad weak peak at 1737.77 cm^{-1} for aromatics combination bands. A broad medium peak at 1634.84 cm^{-1} for aromatic C=C bending, alkenyl C=C stretching and C-C=C symmetric stretch of alkenes. Lastly a broad weak peak at 1365.83 cm^{-1} signify bending of phenol (O-H) and alkanes (C-H), while a peak at 1217.15 cm^{-1} indicates C-O-C stretching of ether. FT-IR spectra obtained for

 DOI: 10.1051/matecconf/20166904002

GSE-OPC are tally with [36], where proanthocyanidins or condensed tannins are polyphenol compounds having complex chemical structure which the oligomers are joined by carbon-carbon bonds. The FT-IR spectra obtained after addition of GSE-OPC during crosslinking stage showed similar pattern to GSE-OPC spectra, indicated that GSE-OPC had coated the cooled capsules droplet of Chitosan-Gelatin (B). This result aligned with [37] which explained that if the oligomers of polyphenol are insoluble in the droplets (which refer to capsules); a membrane is formed around them thus encapsulating the droplet. The polyphenolic compound constitutes the droplet, and the crosslinking reaction stabilizes the molecule which was proven by establishment of ester bonds detected from infrared spectroscopy. Finding from FT-IR spectra in crosslinking stage also confirmed that the decreasing of absorbance intensity at 1635.26 cm^{-1} of N-H bend and increasing of absorbance intensity at 1737.92 cm^{-1} of C-O stretch of ester group that characterized Gel-B was due to crosslinking reaction from polyphenol element in GSE-OPC. FT-IR spectra in harvesting stage which presented in Fig. 4(g) shows no changes compared to previous stage (crosslinking) since harvesting stage just adding few more of water to give extra fluidity hence promotes better settling and separation process of the CO capsules.

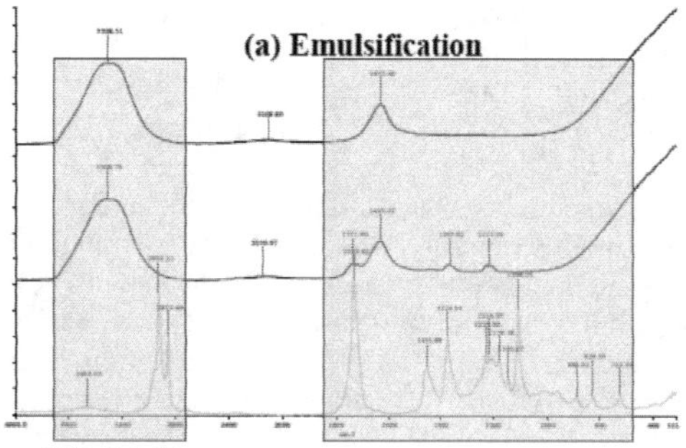

__ CO+Gel-B
__ Gel-B
__ CO

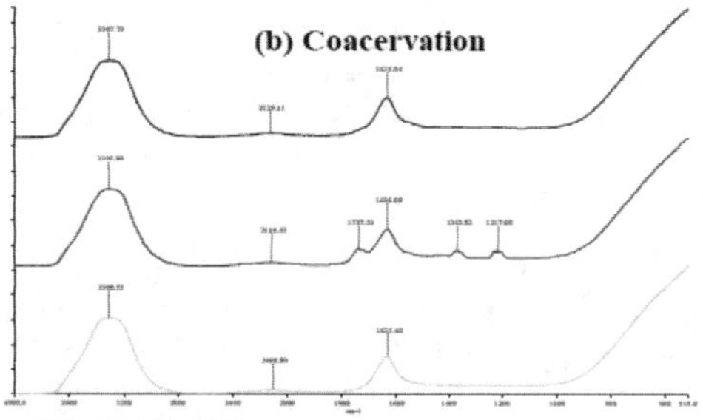

__ CO+Gel-B+Chi
__ Chi
__ CO+Gel-B

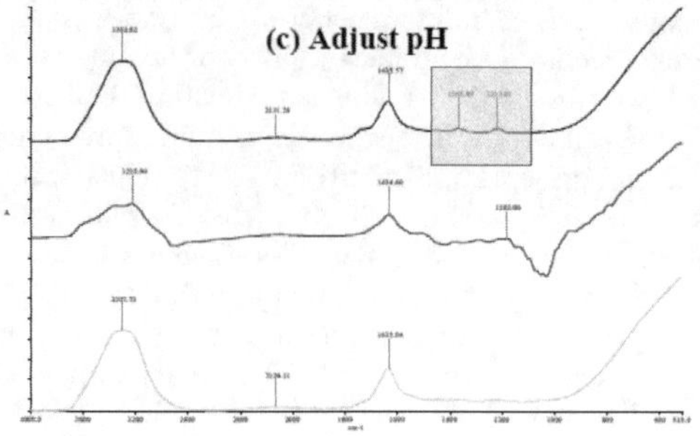

__ CO+Gel-B+Chi (pH 5)
__ NaOH
__ CO+Gel-B+Chi

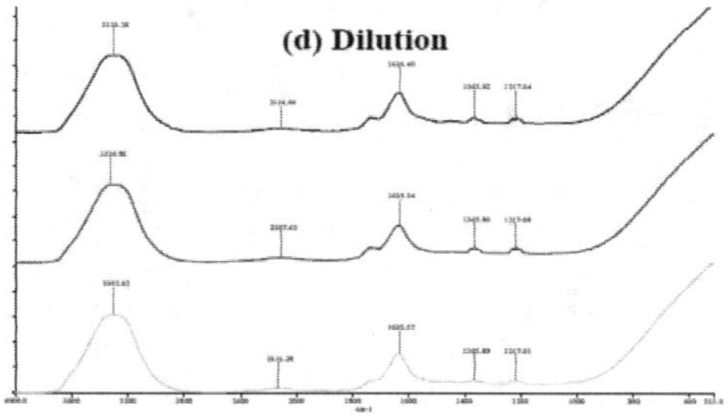

__ CO+Gel-B+Chi (pH 5) dilute
__ DIW (30°C)
__ CO+Gel-B+Chi (pH 5)

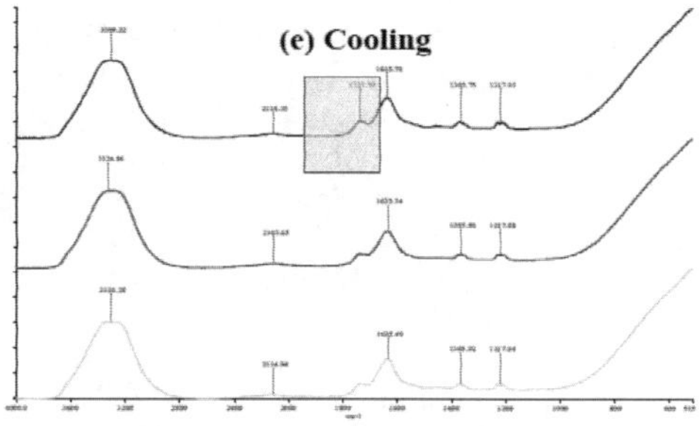

__ CO+Gel-B+Chi (pH 5) dilute (10°C)
__ DIW (10°C)
__ CO+Gel-B+Chi (pH 5) dilute

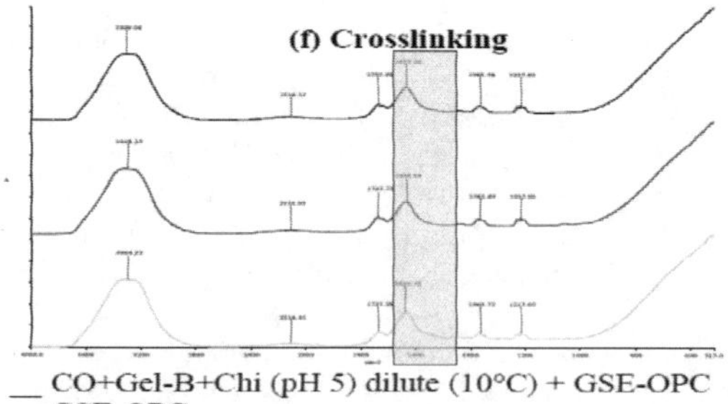

__ CO+Gel-B+Chi (pH 5) dilute (10°C) + GSE-OPC
__ GSE-OPC
__ CO+Gel-B+Chi (pH 5) dilute (10°C)

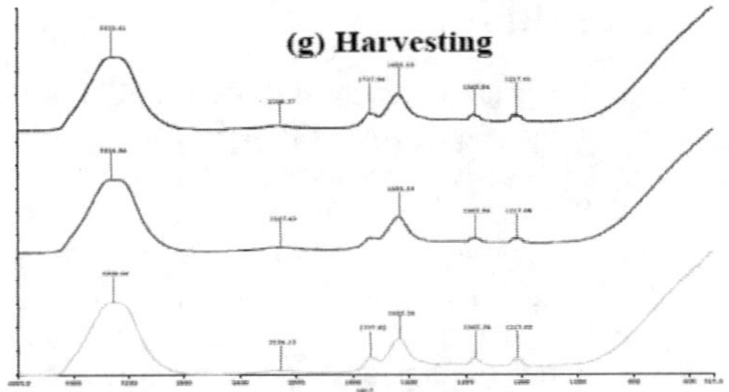

Figure 4. FT-IR spectra for each Citronella oil (CO) encapsulation process from (a) to (g).

Fig. 5 showed FT-IR spectra obtained at CO capsules collection process after settling process for more than 12 hours. Comparing each FT-IR spectra in Fig. 5, we distinguished that each spectra at top, middle and bottom layer showed quite similar pattern to spectra during harvesting stage. However, there are few significant peak noticed in top layer spectra at peak 2924.09 cm^{-1} for alkanes C-H strech, alkanes C-H bend at peak 1456.16 cm^{-1} and alcohol C-O strech at peak 1229.53 cm^{-1} and 1108.82 cm^{-1} respectively. The alcohol and alkanes peak here indicates CO existence at top layer which might be came from the unencapsulated CO accumulated after settling process or from the release of CO capsules at top layer. Since the middle and bottom layer FT-IR spectra showed quite similar pattern to spectra during harvesting stage without any sign of CO characteristics, we can said that middle layer consist of excess polymer with lots of water (dilute) and bottom layer contain concentrate excess polymer due to settling process. This finding was supported by [33], which explained that separation of the turbid reacted solutions in complex coacervation will produce two liquid phases, namely, the polymer-rich phase at the bottom and the supernatant at the top.

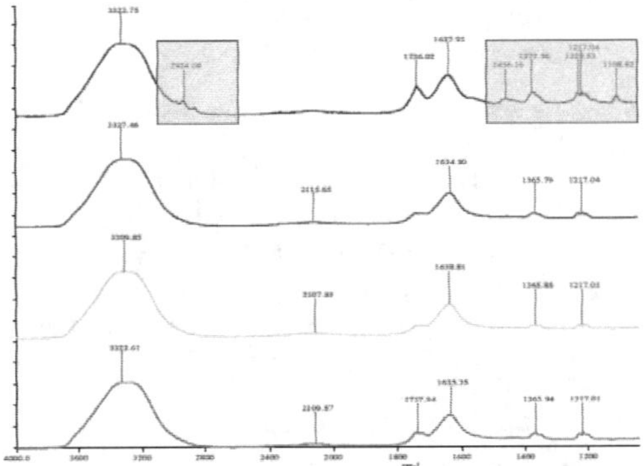

Figure 5. FT-IR spectra at Citronella oil (CO) capsules product collection process.

3.3 Optical microscope image

Fig. 6 shows the optical microscopic image of each process step of Citronella oil (CO) encapsulation conducted from (a) to (g). In emulsification process (a), packed of CO droplets were observed due to high stirring process within transparent appearance of Gel-B solution. Coacervation process (b), takes place when the polycation from Chi solution attract to polyanions from Gel-B which surrounding CO droplet. From Fig. 6(b), the capsules were observed having spherical shape and also in a packed surrounding. After adjusting pH of the process to pH 5 in (c), it was observed that the capsules surface become smoother but a certain degree of agglomeration happens due to the packed process surrounding. After dilution process (d), single spherical shape capsules with smooth surface were observed. The agglomeration occurrence in this process become lower since addition of de-ionized water (DIW) dilute the process system and larger the process surrounding which separate the capsules from each other. Fig. 6(e) shows the capsule condition after cooling process. We claimed here that the wall capsules were hardened since the cooling process was conducted below the gelation temperature of Gel-B which is at 10°C. In crosslinking process (f), there are no major changes in surface morphplogy were observed. The final process of harvesting (g) shows that the capsules are in single spherical shape and having smooth surface. In agreement with [14] based on observation obtained from optical micrograph, gelatin microcapsule appeared to be in round shape with relatively smooth surface. This statement supports earlier observation of CO harvesting process (g), since the operating wall ratio in this study utilize higher amount of gelatin. Thus, it is possible for gelatin to dominate the physical properties of CO capsules obtained.

Fig. 7 shows microscopic images of Citronella oil (CO) capsules surface morphology at top layer after settling process We observed that the final products of CO capsules are in spherical shape which having irregular size. Some of the capsules were attached together causing agglomerate due to the hardened free gelatin-B after cooling until settling process. Agglomeration of capsules during wall formation was a common phenomenon in many microencapsulation processes [38]. As the wall materials change from liquid to solid form, they often went through a sticky stage which makes agglomeration difficult to be avoided [38]. From Fig. 7, we roughly described that the CO capsules obtained in this study are in micrometer size where the diameter varies from 81.63 μm to 156.74 μm with rough surface and almost spherical in shape. This result quite similar with [39] which obtained slightly rough surface yet close to spherical shape as a whole with average particle sizes was 294.7±14.2 μm at emulsification stirring speed of 500 rpm. However, there is quite different where capsule produce in this study are much smaller in size, since [39] applied gelatin-arabic gum system as wall material. This statement was supported by [24] which claimed capsules produced by Chitosan-Gelatin (B) system (30 μm approximate diameter) are much smaller than those produced from Gelatin-Arabic gum system (100 µm approximate diameter). It might be due to the different in viscosity between these two types of complex system where Chitosan-Gelatin (B) system may be less viscous compared to Gelatin-Arabic gum system. Maji and Hussain (2009) revealed that increased in viscosity of the medium resulted in the formation of larger particle droplets during emulsification process. Less viscous material will increase shear rate thus produce smaller particle. This finding was supported by [40] that size of the coacervate suspensions decreased with decrease in viscosity of the material.

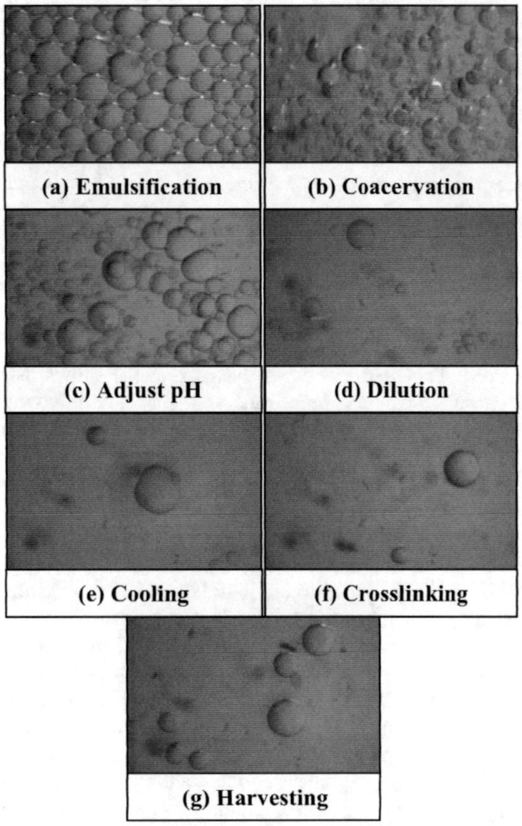

Figure 6. Microscopic image at 30 x magnifications for each Citronella oil (CO) encapsulation process from (a) to (g).

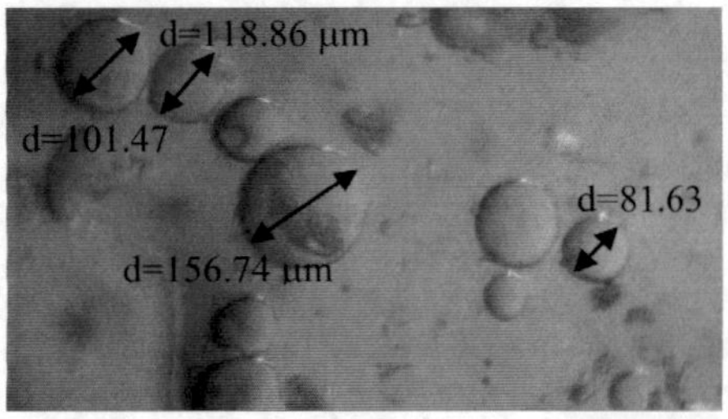

Figure 7. Microscopic images of Citronella oil (CO) capsules surface morphology at top layer after settling process under 30 x magnifications.

4 Conclusions

We found that zeta potential values for Chi acts as polycation within range of pH 4.0 to 5.6, whereas Gel-B acts as polyanion above its isoelectric point (IEP) around pH 4.9. Wall ratio of 1:35 and pH 5 was selected as design operating process for CO capsule production for this study. FT-IR analysis found that Gel-B had coated the CO droplet during emulsification stage, with Chi started to interact with Gel-B to form polyelectrolyte complex in adjust pH stage, CO capsules solidified at cooling process and were hardened by GSE-OPC layer during crosslinking process. Final product of CO capsules after settling process was identified at the top layer. The other middle and bottom layer were indicated as excess polymer with lots of water. Optical micrograph of CO capsules shows nearly round shape with quite smooth surface throughout each process stage. Final product of CO capsules obtained in this study can be described as having diameter varies from 81.63 µm to 156.74 µm with almost spherical in shape with rough surface. Low agglomeration was identified at the final process of CO capsules collection even though crosslinking stage that utilized GSE-OPC as the crosslinker was conducted as the counter measure. More study on crosslinking reaction and application of drying process after CO capsules collection could be suggested for future works. In this study, we can conclude that the proposed method of complex coacervation using chitosan-gelatin (B) system was successfully encapsulated the CO. Further investigation on CO capsules properties and performance will be carried out in the next study. We expect that this study will give better controlled release of CO which will further benefit industry that utilizes CO in their product.

Acknowledgments

The authors fully acknowledged Ministry of Higher Education (MOHE) Malaysia and Universiti Teknologi MARA Malaysia for the financial under research grants of 600-RMI/RAGS 5/3 (76/2013)

References

1. M.H. Hamzah, H. Che Man, H. Jamaludin, Z. Zainal Abidin, International Journal of Chemical Engineering and Applications, **3**, 173-177 (2012)
2. N.A.M. Zin, R. Raslan, J. Jai, I. Subuki, *2nd International Conference on Green Chemical Engineering Technology* (GCET, 2015)
3. N.M. Ali, R. Raslan, J. Jai, I. Subuki, *28th Symposium of Malaysian Chemical Engineers* (SOMChe, 2015)
4. J. Mercola, Citronella oil: the all-natural bug repellent, *Mercola.com*. Retrieved on February 26, 2016 from http://articles.mercola.com/herbal-oils/citronella-oil.aspx
5. R.E.D. Facts: oil of citronella, *United State Environmental Protection Agency*. Retrieved on February 26, 2016 from

http://www3.epa.gov/pesticides/chem_search/reg_actions/reregistration/fs_PC-021901_1-Feb-97.pdf

6. C. Bond, K. Buhl, D. Stone, (2013), Oil of citronella: general fact sheet, *National Pesticide Information Center, Oregon State University Extension Services*. Retrieved on February 26, 2016 from http://npic.orst.edu/factsheets/citronellagen.pdf
7. B. Solomon, F.F. Sahle, T. Gebre-Mariam, K. Asres, R.H.H. Neubert, European Journal of Pharmaceutics and Biopharmaceutics, **80**, 61-66 (2011)
8. M.A. Manaf, J. Jai, A.N. Mustapa, R. Raslan, I. Subuki, Applied Mechanics and Materials, **625**, 860-863 (2014)
9. U. Sakulku, O. Nuchuchua, N. Uawongyart, S. Puttipipatkhachorn, A. Soottitantawat, U. Ruktanonchai, International Journal of Pharmaceutics, **372**, 105-111 (2008)
10. A. Wany, S. Jhaz, V.K. Nigam, D.M. Pandey, International Journal of Advanced Research, **1**, 504-521 (2013)
11. N. Rani, A. Wany, A.S. Vidyarthi, D.M. Pandey, Current Research in Microbiology and Biotechnology, **1**, 98-103 (2013)
12. M.F. Maia, S.J. Moore, Malaria Journal, **10**,1-15 (2011)
13. S.K. Chinta, P.W. Pooja, Indian Journal of Engineering, **3**, 37-40 (2013)
14. S.Y. Cheng, M.C.W. Yuen, C.W. Kan, K.K.L. Cheuk, C.H. Chui, K.H. Lam, International Journal of Molecular Medicine, **24**, 411-419 (2009)
15. F.R.A. Aziz, J. Jai, R. Raslan, I. Subuki, Advanced Material Research, **1113**, 346-351 (2015)
16. L. Li, W. Au, T. Hua, D. Zhao, K. Wong, Textile Research Journal, **0**, 1-6 (2013)
17. X. Jun-xia, Y. Hai-yan, Y. Jian, Food Chemistry, **125**, 1267-1272 (2010)
18. D. Yu, W. Qiao, Q. Li, G. Pei, Journal of Fiber Bioengineering & Informatics, **5**, 67-76 (2012)
19. S.K. Basu, K. Kavitha, M. Rupeshkumar, Journal of Scientia Pharmaceutica, **78**, 79-92 (2009)
20. C.A. Peptu, G. Buhus, M. Popa, A. Perichaud, D. Costin, Journal of Bioactive and Compatible Polymers, **25**, 98-116 (2010)
21. M. Dennis and W. Brownrigg, "Development of micro-encapsulated phase change materials," *Australia and New Zealand Solar Energy Society Conference (Solar 2006)*, ed. Conference Program Committee, Australian and New Zealand Solar Energy Society (ANZSES), Canberra Australia, pp. 1-12, 2006.
22. S. Kim, M.E. Nimni, Z. Yang, B. Han, Wiley InterScience, 442-450 (2005)
23. T.K. Maji, M.R. Hussain, *Journal of Applied Polymer Science*, **111**, 779-785 (2008)
24. A.S. Prata, C.R.F. Grosso, Carbohydrate Polymers, (2014)
25. M.K. Kang, J. Dai, J. Kim, Journal of Industrial and Engineering Chemistry, **18**, 355-359 (2011)
26. H. Alavi Talab, M. Ardjmand, A.A. Motallebi, R. Pourgholam, Iranian Journal of Fisheries Sciences, **9**, 199-208 (2009)
27. M.M.M. Specos, G. Escobar, P. Marino, C. Puggia, M. Victoria, D. Tesoriero, L. Hermida, Journal of Industrial Textiles, **40**, 13-32 (2010)
28. L.F. Siow, C.S. Ong, Journal of Food Processing & Technology, **4**, 1-5 (2012)
29. M.E.I. Badawy, E.I. Rabea, International Journal of Carbohydrate Chemistry, **2011**, 1-29 (2011)
30. M.C. Silva, C.T. Andrade, Polímeros: Ciência e Tecnologia, **19**, 1-5 (2009)
31. K.E. Patrick, S. Abbas, Y. Lv, I.S.B. Ntsama, X. Zhang, Pakistan Journal of Food Sciences, **23**, 17-25 (2013)
32. R.T. Balch, ANALYTICAL EDITION, **3**, 124-125 (1931)
33. H.B. Bohidar, J. Surface Sci. Technol., **24**, 105-124 (2008)
34. M.E. Jason, D.J. Kalota, U.S. Patent 5 540 927, (1996)
35. S.N. Molnes, Physical properties of gelatin based solid emulsions: effects on drug release in the GI tract, *M.S. thesis, Dept. of Biotechnology, Norwegian Univ. of Science and Technology,* Trondheim, (2013)
36. N. Tazzini, (2014), Proanthocyanidins: definition, structure and absorption, *Tuscany Diet.* Retrieved on March 4, 2016 from http://www.tuscany-diet.net/2014/02/12/proanthocyanidins-definition-structure-absorption/
37. A. Munin and F.E. Lévy, Pharmaceutics, **3**, 793-829 (2011)
38. H. Nack, J. Soc. Cosmetic Chemists, **21**, 85-98 (1970)
39. C.P. Chang, T.K. Leung, S.M. Lin, C.C. Hsu, *Colloids and SurfacesB: Biointerfaces,* **50**, 136-140 (2006)
40. E. Kizilay, A.B. Kayitmazer, P.L. Dubin, Advances in Colloid and Interface Science, **167**, 24-37 (2011)

MATEC Web of Conferences **69**, 04003 (2016) DOI: 10.1051/matecconf/20166904003
ICCPE 2016

A comparison between the effects of sepiolite and silica on mechanical properties and thermal stability of NR/EPDM blend

Nattawat Winya[1,2] and Nanthiya Hansupalak[2]

[1]*Defence Technology Institute (Public Organization) Ministry of Defence, Nonthaburi Thailand*
[2]*Department of Chemical Engineering, Kasetsart University, Bangkok, Thailand*

Abstract. This study focused on improving the ablation properties and mechanical properties of natural rubber (NR)/ethylene propylene diene monomer (EPDM) blends by using two different commercial fillers: sepiolite and silica. The results showed that sepiolite and silica similarly improved both properties in the loading range of interest (0-12.5 phr). However, sepiolite was superior than silica as using less loading to acheive comparable mechanical and thermal properties. The optimum loading of sepiolite was 5 phr, while that of silica was 10 phr. For both filler systems, the minimum ablation rate found was approximately 0.17 mm/s, and the tensile strength and the hardness of corresponding NR/EPDM composite were improved after modification with sepiolite or silica.Thus, the composite reinforced with sepiolite or silica is suitable for many applications, such as internal insulations of rocket motor.

1 Introduction

The insulation for applications such as in rocket motor have very demanding requirement. Internal insulations in rocket motor protect structures from heat flux produces by the combustion of propellant in motor case. The combustions of propellant produces high velocity gases with temperature more than 2000°C. The elastomers used for solid rocket motors (SRM), ethylene propylene diene monomer rubber (EPDM), nitrile butadiene rubber (NBR) and silicone rubber are used for the internal thermal insulation in solid rocket motor (SRM) [1, 2]. The development of novel materials on a commercial scale to fulfill these complicated requirements of high temperature internal insulations for rocket motor and development of further uses of natural rubber has been applied for the insulation of rocket motor. Although, fillers have been successfully that the addition of reinforcing fillers used to enhance the mechanical properties of the composites if properly added [3]. However, fillers are available in diverse range of properties such as specific gravity, color, electrical and thermal conductivities etc. therefore; desired properties of composite can be very easily varied by selecting proper filler. Silica is an amorphous material composed of silicon and oxygen atoms connected with silanol (Si-OH) both inside and on the surface. The silanols group that provide hydrophilic characteristics to silica surface and moisture readily on silica surface [4].

We would like to note that during the mixing process, the chemical reaction occurs between aggregates of silica and natural rubber chains. To avoid the complications, a silane coupling agent is needed to improve silica-rubber interaction and silica dispersion as well as to prevent accelerator adsorption on the silica surface [5-9]. Sepiolite is a hydrated magnesium silicate clay (Mg2H2SiO3O9.xH2O), Sepiolite has a similar structure to the layered structure of montmorillonite except that the layers lack continuous octahedral sheet [10-11]. The acicular morphology of sepiolite generally improves adhesion/compatibility with polymeric matrices and provides an excellent reinforcing effect on polymer [12] and sepiolite improved the endothermic nature of decomposition reaction. This filler increased the tendency of the matrix to charring. Thus, the sepiolite can be selectd as promising materials for innovation of ablative materials [13]. The oservations confirmed the presence of Needles of sepiolite shown that Fig. 1.

Figure 1. SEM micrographs of sepiolite powders.

Lombardi et al. focused on the thermal properties of poly(phenylene oxide)/polystyrene (PPO/PS) composites filled with modified clay, hydrotalcite and Sepiolite. The hydrotalcite added PPO composites shows that the degradation and combustion occurs at low temperatures and in a shorter time than the pure PPO polymer. While the modified clay added PPO composites and the Sepiolite added PPO composite were a higher thermal stability and a more endothermic decomposition, but the modified clay added PPO composites was brittle char layer during combustion. Thus, the Sepiolite added PPO composite can be selected as promising materials for innovative ablative shields [13]. Yue Guan et al. studied the ablative properties and ablation mechanisms of hydrogenated nitrile butadiene rubber (HNBR) composites filled with fumed silica (FS), organically modified montmorillonite (OMMT), and expanded graphite (EG).From studies HNBR composite as a result found that the ablative of HNBR/OMMT composites exhibits the best ablation resistance among three kinds of HNBR based composites [14]. Jovanović et al. studied the rheological, mechanical properties, thermal stability and morphology of acrylonitrile-butadiene/ethylene-propylene-diene (NBR/EPDM) rubber blend with carbon black (CB) filler ranging from 60 to 100 phr. The experimental results showed that the optimum cure time (tc90), tensile strength and hardness increase but scorch time (ts2) and cure rate index (CRI) decrease with an increase in the CB content in NBR/EPDM rubber blend composite because CB can be attributed to the crosslink density of rubber blend. The NBR/EPDM/CB (80/20/70) rubber blend has maximum tensile strength. However, the composite was degraded at a lower temperature, compared to the NBR and EPDM rubber. Furthermore, it was found that rubber and CB are compatible [15, 16]. The objective of this work was to understand the effects of fillers loading on ablation and mechanical properties of natural rubber (NR)/ethylene propylene diene monomer (EPDM) blends. The influence of two types of filler (i.e.,sepiolite and silica) on the selective ablation and mechanical properties of NR/EPDM-filler composite.

2 Materials & methods

2.1 Materials

NR (natural rubber, STR 5L), Silica (VN-3), Sepiolite (sigma-aldrich), Fiber (Aramid-pulp), Antioxidant (Winstay S), Flame retardant (Dechlorane Puls 515), Antimony Oxide, stearic acid, Silane, Sulphur and oil.

2.2 Measurements

The mechanical properties were measured by tensile TECH 91202 serial no.TBD-2730 according to JIS K6251.

The hardness (Shore D) was measured by Teclock GS-706N according to JIS K6268.

The density measured by METTLER TOLEDO (XS204) according to JIS K6268.

The curing times (t90) of the compounds were obtained by EKTRON at 170 °C.

The compounds were subsequently compression molded by GOTECH (GT-7016) at 170 °C, beased on the respective t90 values.

Ablation rate was measured using oxygen and acetylene ablation testing equipment according to ASTM E285 (Fig. 2). The test specimen shall be a square five replicates and calculate the ablation rate to the following equation is ablation rate = thickness/ablation time.

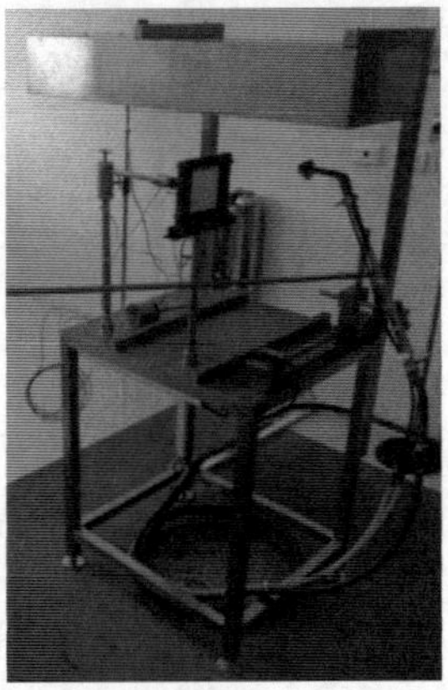

Figure 2. Test specimen of ablation test.

2.3 Preparation of NR/EPDM composite

Preparation of compound formulations are natural rubber was mixed at a temperature of 50 °C. Non-vulcaized natural rubber composite containing antimony oxide, flame retardant, stearic acid, silica, sepiolite and silane coupling agents given in Table 1. The composite rubber investigated contains 0, 2.5, 5, 7.5, 10 and 12.5 phr of type of filler varied, while all the other ingredients remained constant. After that will study curing characteristics, mechanical properties and ablation properties.

Table 1. Formulation for NR/EPDM composite

Raw material and ingredients	(phr) [a]
NR	80
EPDM	20
Aramid-pulp	20
Antioxidant	2
Flame Retardant	40
Antimony Oxide	20
Silica/Sepiolite	0,2.5,5,7.5,10,12.5
Stearic acid	1
Silane	1
Co-agent	2
DCP-40	5

[a] phr stands for parts per hundred parts of rubber

3 Results and discussion

Table 2 shows that the curing characteristics of all rubber composilte with sepiolite and silica samples. The experimental results show that amount of sepiolite and silica have no significant effect on M_L (minimum torque),

t_{s2} (scorch time) and t_{90} (optimum cure time). The table also shows the curing characteristics of the hard rubber under investigation. Regular variations in the maximum torque (M_H) and optimum curing time (t_{c90}) were observed for the reinforced blends. The NR/EPDM with 2.5 phr of sepiolite demonstrated an optimal curing time. The initial decrease in torque.

Fig. 3 displays the maximum torque (M_H) of all rubber composilte with sepiolite and silica samples. It was interesting to find out the maximum torque not much different when sepiolite and silica increased in rubber compound However, the maximum torque values of all the samples are increase of silica and sepiolite loading. The torque then increased due to the crosslinking between the molecular chains. Crosslinks are formed between the unsaturated sites and molecules during vulcanization. The maximum torque increases proportional to the indicating crosslinking density enhancement because of the reinforcement.

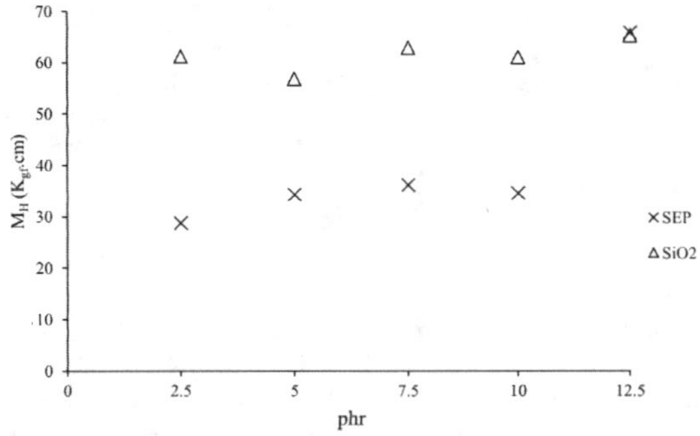

Figure 3. Maximum torque (MH) of hard rubber compounds

Table 2. Curing characteristics of NR/EPDM

	ML	MH	T_{s2}	T_{c90}	CRI
0 SEP	51.04	25.89	0.42	5.19	21
2.5 SEP	55.64	28.71	0.42	5.21	20.9
5-SEP	60.73	34.3	0.42	5.11	21.3
7.5 SEP	63.09	36.14	0.42	5.04	21.6
10-SEP	63.23	34.53	0.42	4.45	24.8
12.5 SEP	35.9	65.7	0.42	4.54	24.3
2.5 Si	33.01	61.11	0.03	5.17	19.5
5 Si	31.7	56.8	0.03	5.02	20
7.5 Si	33.24	62.79	0.03	5.1	19.7
10 Si	32.5	60.95	0.03	5.16	19.5
12.5 Si	35.78	65.18	0.03	5.05	19.9

*the cure rate index (CRI)=100/(t_{90}-t_2)

The effect of sepiolite and silica loading on tensile strength of NR/EPDM blends composite is shown in Fig. 4. Result showed that tensile of vulcanizates increased with the increase of sepiolite and silica loadings. However, the effect was more significant when sepiolite and silica content was up to 7.5 phr and then little decrease at 10 phr for silica. At higher concentration, silica formed aggregates which had negative effect on mechanical properties of the vulcanizates.

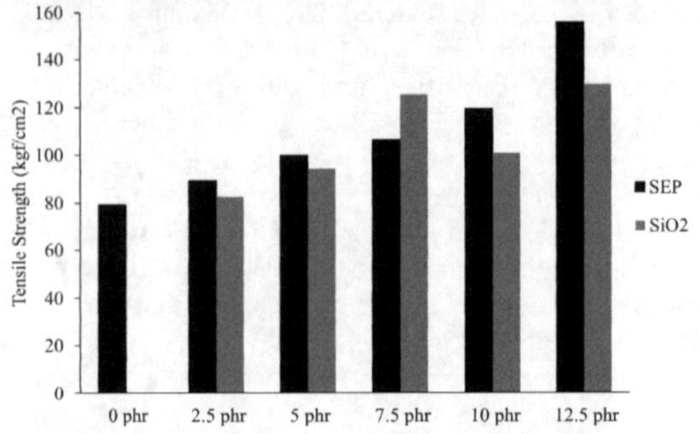

Figure 4. Tensile strength of NR/EPDM blends filled with sepiolite or silica

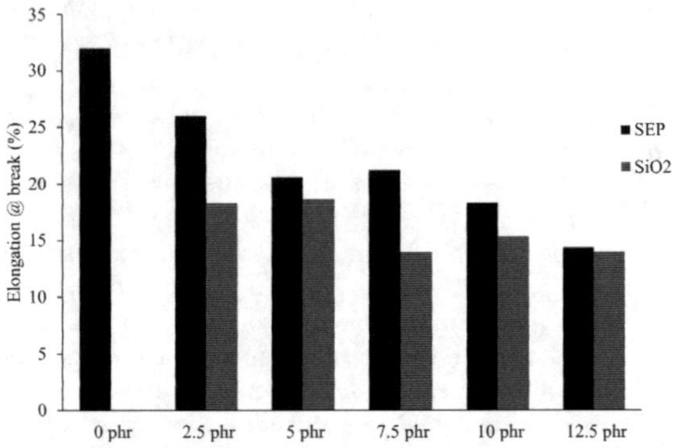

Figure 5. The elongation of NR/EPDM blends filled with sepiolite or silica

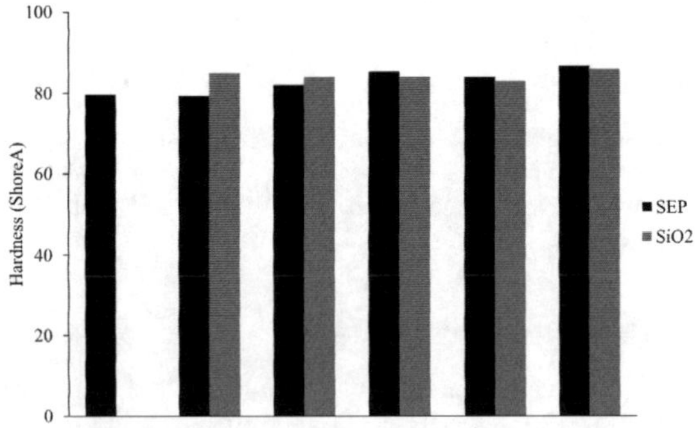

Figure 6. Hardness of NR/EPDM blends filled with sepiolite or silica

The effect of sepilolite and silica loading on elongation of NR/EPDM composite shown in Fig. 5. The result showed that elongation decreased with the increase of sepiolite and silica loading because of the reinforcement.

The Fig. 6. showed the effect of sepiolite and silica loading on hardness of NR/EPDM blends composite

DOI: 10.1051/matecconf/20166904003

based vulcanizates. It clear from result that hardness of the vulcanizates increased with the increase of sepiolite and silica loading. In this study, the ablative properties of NR/EPDM blends filled with sepiolite and silica shown in Fig. 7. The experimental results showed that the ablation properties of NR/EPDM composite were improved after modification with sepiolite or silica. The effect of sepiolite loading on ablation rate composite decreased with increasing of sepiolite loading content in NR/EPDM composite. The NR/EPDM/SEP (80/20/5) rubber blend has minimum ablation rate. The NR/EPDM added SiO2 composites exhibits the best ablation resistance at 12.5 phr of silica content in the NR/EPDM composite.

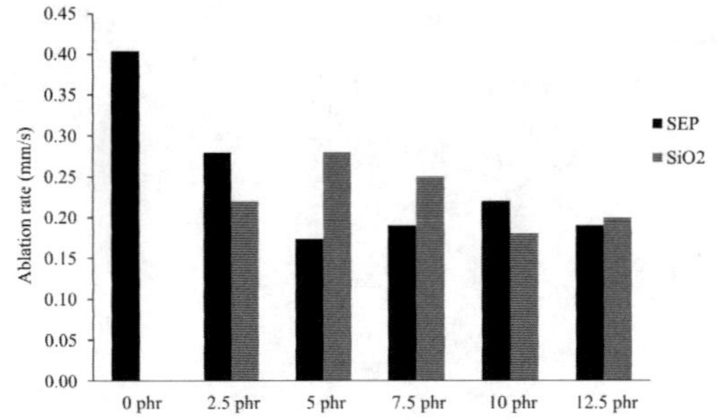

Figure 7. Ablation rate of NR/EPDM blends filled with sepiolite or silica

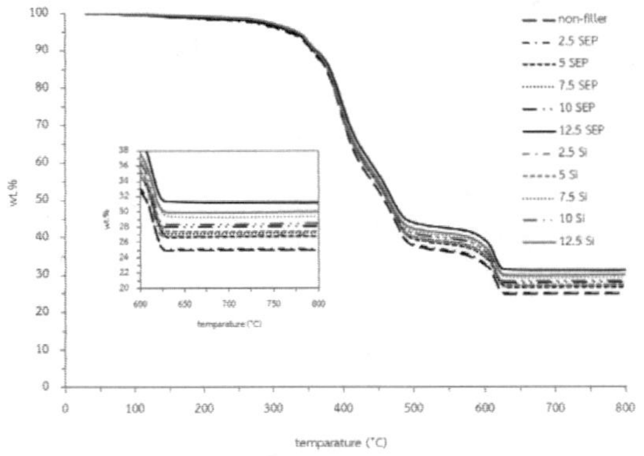

Figure 8. TG curves of NR/EPDM composites

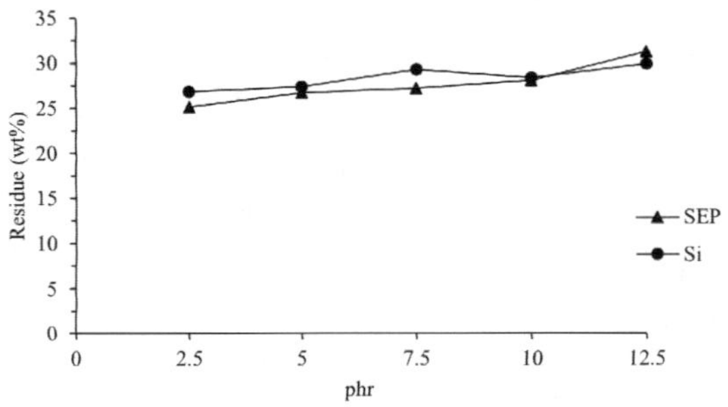

Figure 9. Experiment values of residues at 800 °C for NR/EPDM composite.

The TG thermos-gram of all NR/EPDM composites showed two thermal degradation transitions (Fig. 8.). The 1st transition at 360-450 °C of organic compounds with high molecular weight and second step, 600-635 °C and at 800 °C, the char residue of NR/EPDM composite with 12.5 phr of SEP was 31.3 wt% and NR/EPDM composite with 12.5 phr of silica composite 29.9 wt% (Fig. 9). The sepiolite and silica did not change the degradation of NR/EPDM composites.

Morphological images of ablation test of composites shown in Fig. 10. Fig. 10c showed that char layer of NR/EPDM composites with sepiolite. It can see char layer formed on the surface of the 5 phr of sepiolite was almost continuous and covered the surface. Fig. 10b was surface after ablation test that displays the many sphere structures can be observed on the char layer surface should be made of silica. Sepiolite and silica to NR/EPDM composite enhance thermal ablation performance by forming char layer on the surface of composite.

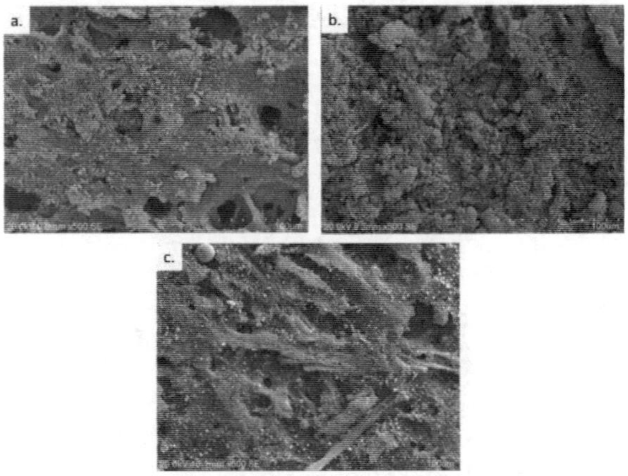

Figure 10. Images of morphology of surface after ablation test (a) NR/EPDM without fiilers, (b) NR/EPDM with 5 phr of SEP and (c) NR/EPDM with 10 phr of Silica

4 Conclusion

The effect of sepiolite and silica loading on tensile strength of NR/EPDM blends composite is showed that tensile of vulcanizates increased with the increase of sepiolite and silica loadings. However, the effect was more significant when sepiolite and silica content was up to 5 phr and then little decrease at 10 phr for silica. At higher concentration, silica formed aggregates which had negative effect on mechanical properties of the vulcanizates. The ablative properties of NR/EPDM blends filled with sepiolite and silica. The experimental results showed that the ablation properties of NR/EPDM composite were improved after modification with sepiolite or silica

Acknowledgment

This work was supported in part by Research and Development Department, The Defense Technology Institute (Public Organization), Nonthaburi, Thailand.

References

1. M.B Khan, Polymer Plast Tech Eng **36** (1996)
2. J.B Tan, World rubber industry(in chinese) **33**, 6 (2006)
3. Q. Quifu Liu, Y. Zhang Xu, H., APPL CLAY SCI **42** (2008)
4. P. Sarawut and R. Nittaya, Polym Test **31**, 5 (2012)
5. Z.Guozhang, S. Liyi, Z Dengsong, F. Xin, Y. Shuai and Z. Jing, Materials & Design **35**(2012)
6. S. Prasertsri andN. Rattanasom, Polym Test **30**, 5 (2011)
7. T.H. Mokhothu, A.S. Luyt, M. Messori, Polym Test **33** (2014)
8. R.V.Sekharan, B.T.Abraham and E.T. Thachil, Materials & Design, **40** (2014)
9. Y. Jiang, X. Zhang, J. HE, L.Yu and R. Yang, Polym Degrad Stabil, **96** (2011)
10. P. ArandaR, A. Kun, Martín-Luengo, S. Letaïef, I. Dekany and E. Ruiz-Hitzky, Chem. Mater. **20** (2008)
11. R. Benlikaya, M. Alkan and I. Kaya, Polym. Compos. **30**(2009)
12. L. Bokobza, A. Burr, G. Garnaud, M.Y. Perrin and S. Pagnotta, Polym. Int. **53** (2004)
13. M. Lombardi, P. Fino, G. Malucelli, and L.Montanaro, Composite Structures, **94** (2012)
14. Y. Guan, Z. Ling-Xi, Z. Li-Qun, L.Yong-Lai, Polym Degrad Stabil. **96** (2011)
15. V. Jovanović, S. Samaržija-Jovanović, J. Budinski-Simendć, G. Marković, M. Marinović-Cincović,, Composites: Part B **45** (2013)
16. S. Samaržija-Jovanović, V. Jovanović, G. Marković, S. Konstantinović, M. Marinović-Cincović, Composites:Part B
42 (2011)

MATEC Web of Conferences **69**, 05001 (2016) DOI: 10.1051/matecconf/20166905001
ICCPE 2016

The effectiveness Study of Different Membranes in Treating Industrial Wastewater

Munawar Zaman Shahruddin, Izzah Sofiahanun Binti Ishak, Nur Hidayati Othman, Nur Hashimah Alias and Nur Azrini Ramlee

Universiti Teknologi MARA, Malaysia

Abstract. This paper focuses on the effectiveness study of different membranes in treating petroleum based wastewater. Polysulfone (PSF) and Polyether Sulfone (PES) were prepared using phase inversion method in NMP solvent. The composition of membrane was varied by manipulating the PVP additives concentration (5 wt%) in the membrane dope solution. The morphological study of prepared membranes was analysed using Field Emission Scanning Electron Microscope (FESEM) while the permeating flux and Chemical Oxygen Demand (COD) rejection rate were determined accordingly by using the refinery wastewater as the feed. The result found that the permeating flux increased as the operating pressure increased. However, for COD rejection, it was reported that only certain membrane show the best rejection. To clarify and support this finding, the membrane performance was further analyzed by determining the Total Dissolved Solid (TDS) rejection for PSF2 and PES2 membrane samples. The result showed that both membranes recorded relatively high TDS rejection which can be justified by the morphological studies of each samples. Therefore, this study proved that the membrane treatment have a high potential to treat the industrial wastewater for a larger scale.

1 Introduction

Ever since the establishment of the Petroleum Development Act (1974), it can be seen that development of oil and gas industry in Malaysia has been increasing year by year. This situation brings a lot of benefits and advantages in terms of country's profit, supporting a huge of job vacancies to local people and also gives a lot of investment opportunity to foreign and local investor.

However, this increasing industrialization has given some impact on ecosystem and environment which human relies on. This is because the wastewater from industry and refinery has become one of major source that contributes to water pollution.

Discharging industrial wastewater without being treated further may contaminate water and soil that eventually will result-in disturbances to the ecosystem. On initial discharge, these wastewaters may comprise high levels of inorganic pollutants which can be easily biodegradable, however Total Dissolved Solids (TDS), Biochemical Oxygen Demand (BOD) or Chemical Oxygen Demand (COD) can give bad impact on the ecosystems if their content in the waste water even maybe in tens of thousands mg/L.

The conventional wastewater treatment consists of a combination of physical, chemical and biological process and work to remove solids, organic matter and sometimes, nutrients from wastewater. It starts with the preliminary processes, followed by primary treatments which are focusing the physical separation of contaminants in the effluent. Next step would be secondary treatment which involves biological degradation of bacteria in order to decompose the organic pollutants in the effluent. The last stage of treatment would be tertiary treatment which is additional treatment if both of earlier treatments cannot achieve the standard or requirement as per requested.

In a conventional wastewater treatment plant operation, three major process mainly involve: biochemical treatment, liquid/solid separation operation for sludge treatment and valorization [1]. In scope of petroleum industry, there are two basic treatment stage involve where are mechanical and physicochemical treatment in the first stage and followed by the advanced treatment of the pre-treated primary effluent [2].

Apart of that, there are other several methods that been proposed in treating industry wastewater such as electrochemical coagulation [3], electron Fenton [4], membrane bioreactor [5], photocatalytic reaction [6], catalytic ozonation [7] and membrane [8]-[10].

A lot of study on membrane technology has been conducted as the researchers finds that membrane could be a good option to treat and improve the quality of wastewater. According to Saljoughi & Mousavi [10], membrane processes have been regarded as an advanced treatment compared to conventional wastewater treatment processes. The benefits of advanced treatment compared to conventional process are due to its characteristic such as selective preparation, continuous and automatic operation, and easy and well-arranged process conduction.

There are some researches done by several experts on the application of microfiltration or ultrafiltration membrane in industrial wastewater treatment. A review on the applicability of integrated/hybrid membrane processes in water treatment and reverse osmosis desalination plants have been studied where the result founds that it has huge potential to be widely applied and can lead to breakthrough in solving the problem of water scarcity [11].

Salahi et al. [12] reported the performance evaluation of five different polymeric membranes including two ultrafiltration and three ultrafiltration for treatment of an industrial oily wastewater, where the result found that ultrafiltration membrane resulted-in highest flux and higher rejection. Lin & Lan [8] investigated the treatment of waste drawing oil which is commonly used in the cable and wire industries by employing semi batch ultrafiltration (UF) and reverse osmosis (RO). The result showed that the water quality of permeated from combine processes displayed excellent performance. Salahi et al. [13] tested the asymmetric Polyether Sulfone ultrafiltration membrane for oil wastewater treatment and the result reported that Polyether Sulfone is effective in removing COD, turbidity etc.

Chakrabarty et al. [9] in their study found that morphological properties of membrane due to addition of PVP and PEG have a significant influence on the permeate flow rate and oil removal. This will lead to the emergence of the wastewater treatment by using the polymeric membranes.

2 Methodology

2.1 Materials

The main polymers used in this study are PSF and PES with having the same average weight of 35 000 Da were provided by BASF. Reagents grade N-methyl-2-pyrrolidone (NMP) (99.5% purity) used as solvent were purchased from Anachemia Ltd. Reagent grade Polyvinyl pyrrolidone (PVP) (average molecular 58 000 Da) was used as additive in casting solution in order to get membrane with varying morphological and permeation properties. Distilled water was used as main non solvent in the coagulation bath in preparing the membrane. Wastewater sample was collected from the refinery plant.

Table 1. Membrane dope composition

Polymer/Membrane (20 wt%)	PVP (wt%)	NMP (wt %)
PSF1	0	80
PSF2	5	75
PES1	0	80
PES2	5	75

2.2 Membrane preparation

PSF and PES beads were dried in oven for 1 hour at 80 °C before being used. Phase inversion method was used to synthesis the flat sheet membrane by using distilled water as non-solvent. Both of the polymers were dissolved separately in the solvent of NMP at room temperature (25 °C). After the polymer was completely dissolved in NMP solution, then PVP reagent was added. Table 1 represent the composition of different membranes. The PSF and PES concentration was kept constant at 20%, keeping the solvent and additive concentration at 80%. Each dope solution needs to be mixed properly by stirring the solution at 200 rpm. Next, the prepared solution was casting manually using casting knife with the thickness of 150μm. The synthesized membrane was then immersed in distilled water for 24 hours. Then the formed membrane was dried at room temperature for one day. The synthesized membranes were designated as PSF1, PSF2, PES1 and PES2.

2.3 Membrane characterization

2.3.1 Permeating flux and rejection rate

In this study, permeating flux, J ($m^3/m^2.s$) for each membrane was determined at different pressure (2 and 4 Bar) using wastewater as the feed by employing the following equation:

$$J=Q/(A.t) \tag{1}$$

where Q is the volume of permeate (m^3), A is the effective membrane area (m^2) and t is the time obtain the volume of Q (s).

To assess the separation efficiency of each membrane, the rejection rate, R (%) is used and can be calculated using the following formula:

$$R= (1- C_f/C_p) \times 100 \tag{2}$$

where C_f and C_p are the concentration (any parameter of interest) of feed and membrane permeate. The membrane setup can be found in Fig. 1.

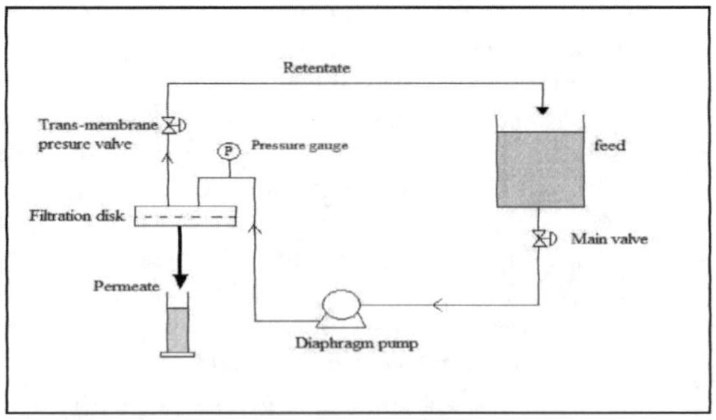

Figure 1. Membrane permeation testing rig

2.3.2 Analytical COD and TDS

Analysis on the COD concentration of the feed and permeate sample were conducted using Hach Spechtrophotometer (Model DR2800). Meanwhile TDS meter (Model Hach) was used to determine the concentration of TDS in feed and permeate.

2.3.3 Morphological study

The prepared membranes were characterized by morphological analysis and permeation experiments. The morphological of the prepared membrane was studied using FESEM. The membrane sample was dried and then fractured cryogenically in liquid nitrogen before mounting on sample stud. The samples were then sputtered with a thin layer of gold using a sputtering apparatus. Various FESEM images were taken at magnification of 1500 for cross sectional views.

3 Results and discussion

The refinery wastewater was characterized in terms of several water quality parameters as mentioned in the methodology before being treated. The value for all parameters can be seen in Table 2.

Table 2. Refinery wastewater sample characteristics

Wastewater Characteristic	Value
Conductivity	3.38 mS/cm
Total Dissolved Solid (TDS)	1758 mg/L
Chemical Oxygen Demand (COD)	153 mg/L

3.1 Membrane permeating flux

Fig. 2 shows the trend of permeating flux with two different operating pressures for four different types of membrane.

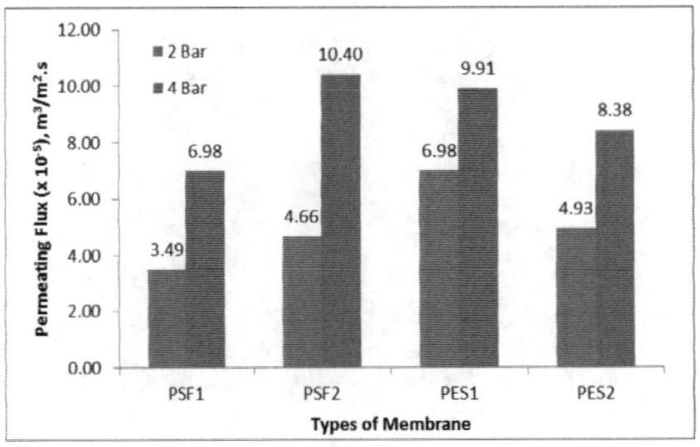

Figure 2. Permeating flux for all membrane samples

As can be seen in the figure, the trend shows that the flux increase linearly with the operating pressure for all membrane sample. Specifically, for PSF membrane (PSF1 and PSF2), the graph shows that membrane fluxes for PSF2 is higher than PSF1. This shows that the addition of PVP additive affect the membrane performance by increasing the flux value, hence making the membrane performance become well. This result is coherent with the study done by Pagidi et al.[14], where their finding shows that all the blended membrane exhibited a higher permeation flux than the pure membrane. For PES, the graph also show the same trends where both of PES1 and PES2 have increasing flux when operating pressure increase. In term of additive addition into PES membrane solution, the trend recorded vice-versa as to compare with PSF membrane. This phenomenon is resulted from the non-uniform distribution of micropores as a result of the addition of PVP in the membrane dope solution. Ismail & Hassan [15] also found the same finding in their study. Plus, this result could be strongly supported with Barth et al.[16] where their study also shows the same result.

From the graph, a comparison for PSF and PES membrane shows that the fluxes are higher for PSF membrane. The reason for this finding could be credited to the thick sponge-like support layer of the PSF membrane which additionally permits the flux of bigger molecules. Besides, the different polarities of membrane material also could affect the membrane fluxes [16].

3.2 Rejection rate

3.2.1 COD rejection rate

The performance of membrane sample further analyzed by estimating the rejection during permeation experiment with wastewater as feed. For this purpose, the wastewater quality chosen is COD as the indicator to observe the quality of wastewater before and after membrane filtration.

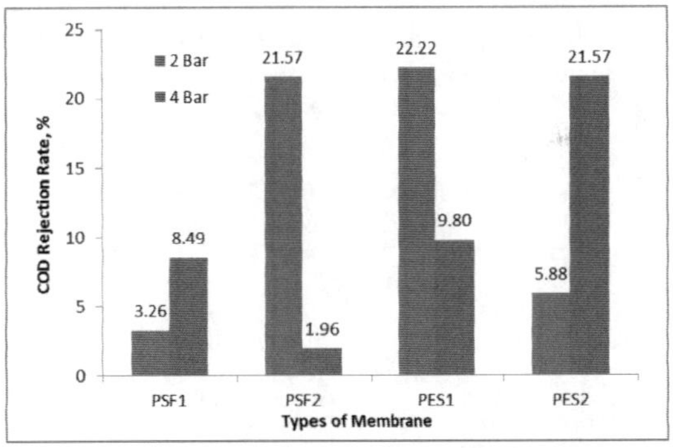

Figure 3. COD rejection rate for all membrane samples

Fig. 3 shows the trends of COD rejection for PSF1, PSF2, PES1 and PES2 membrane sample. As for PSF1, it can obviously be seen that as the operating pressure increased, the COD rejection also increased. This is because as the operating pressure increase, the flux also increase and the amount of retained organic material that contained in wastewater, attached on membrane surface also increase. So the increase of the operating pressure make it possible to decrease the COD concentration in permeate.

While for PES1, the trend shows that COD rejection decline with the increasing operating pressure. If referring to membrane flux (Fig. 2) for PES1, it exhibits that the membrane flux increase with the operating pressure, so theoretically, the COD rejection also should incline with the operating pressure. However, it happen vice versa. This same situation happen to Zheng et al.[17] where their research also show the same result. It can be concluded that the increase of permeate flux, the increase of the amount of the retained organic material across the membrane, thereby resulting an increase in COD concentration in permeate and reduce the COD rejection rates.

As for PSF1 and PSF2, their rejection performance can be compared in terms of additive contribution. As mentioned before, PSF1 is pure polysulfone membrane while PSF2 is the blend membrane with 5 wt% PVP. For comparison, as the operating pressure increase, COD rejection rate for PSF1 increasing, while for PSF2 it is recorded there is declining in COD rejection rate. This is because the roles of PVP additives in polymer solution as a micropores promoter and increase the porosity and supressed the macrovoids formation. So in this study, it can be concluded that as the PVP additives is added in the solution, the pore area per unit surface area (porosity) increase, hence the permeate flux also increase and thereby the COD rejection is decline [18, 19]. This is because as bigger the pore area, more organic material can pass through the membrane.

While for both PES1 and PES2, Fig. 3 shows different trends of COD rejection. As the operating pressure increase, the COD rejection for PES1 decreasing, whereas PES2 show increasing trends. Theoretically, as additive is added, the membrane pore radius increase. Hence, as the operating pressure increase, so the membrane flux also increase, hence causing the rejection rate decrease. The result for PES2 logical accordingly with Ismail & Hassan [15] research.

3.2.2 TDS rejection rate

The performance of membrane is further analyzed with reference of rejection of Total Dissolved Solid (TDS) in the permeate.

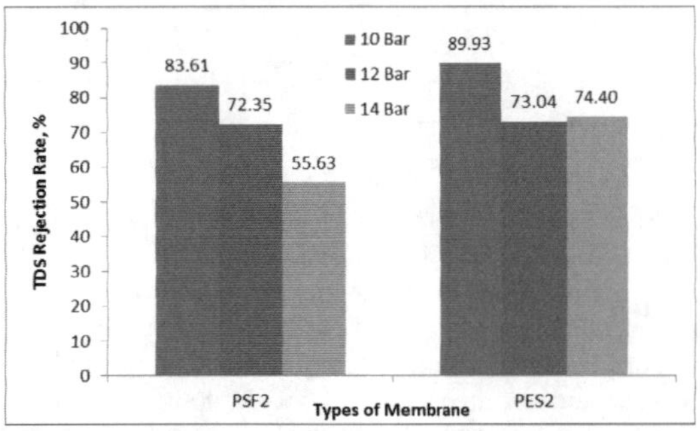

Figure 4. TDS rejection rate for the membrane with the addition of PVP

This analysis is done to clarify and support the result obtained earlier on COD rejection. For this purpose, membrane PSF2 and PES2 are chosen in the analysis in order to study the significance of PVP additive in membrane performance in term of TDS rejection.

For PES2 it is observed that as the operating pressure increase, the rejection of TDS decrease but at 14 bar, it increase slightly from 73% to 74%. This trend does not follow the theory. This situation could arise due to the compaction factor, where it is related to the structure of the membrane especially membrane sub layer. The higher compaction factor, the more likely membrane compact because of the existence of large number of macro voids in the sub layer resulting in narrower passage for permeate flow. Generally, membrane with bigger pore size combined with loose structure result in poor separation performance [9].

3.3 Membrane morphological study

From Fig. 5 it can be seen that membrane formed are having asymmetric structure layer consisting of a dense top layer and a porous sublayer. For comparison, the macropores for PSF2 is smaller compare to PSF1. This is truly true with the fact that PVP additives function to suppress the macropores (Han & Nam, 2002). Therefore the flux data, showing correlation with PVP content in membrane.

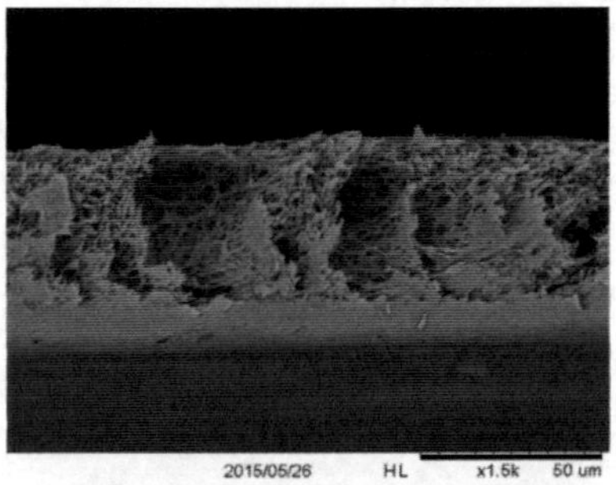

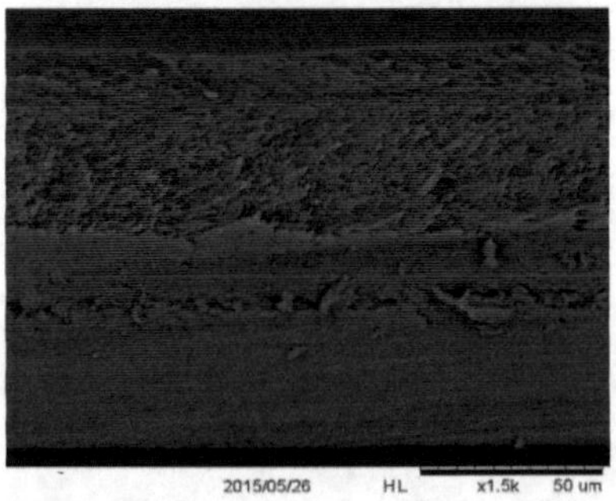

Figure 5. FESEM images of cross sectional for PSF1 (below) and PSF2 (above) membrane sample

As shown in Fig. 6, both PES1 and PES2 have asymmetric structure consists of a dense top layer and finger like porous structure. The cross section image revealed that PES2 have thicker effective skin layer with larger pore size compare to PES1, thus exhibiting low permeating flux and higher rejection rate. This finding correlates with the permeability performance of both membrane as being discussed previous.

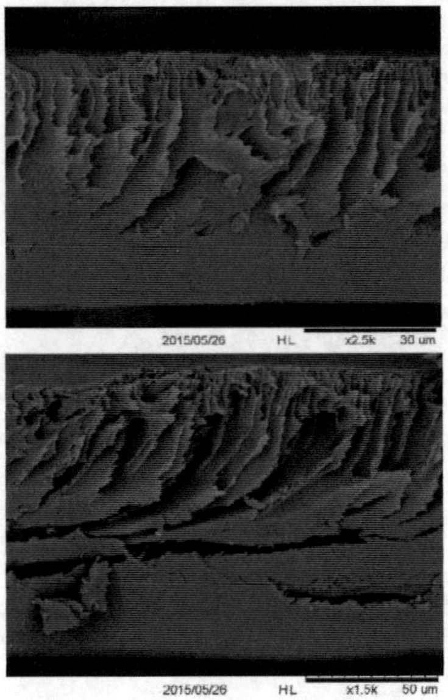

Figure 6. SEM images of cross sectional for PES1 (above) and PES2 (below) membrane sample.

4 Conclusion

The objective of this study is to develop the effectiveness of two different membranes consists of PSF and PES, by manipulating the additive concentration in dope solution to treat the wastewater. Then, the performances of membranes prepared were characterized in terms of permeating flux, COD and TDS rejection and membrane morphological. The following conclusion can be drawn from the study:

1. Membrane have been successfully fabricated using phase inversion method using PSF and PES as the main polymer, NMP as the solvent and PVP as the additive.
2. Membrane testing using wastewater as the feed shows that, as the operating pressure increase, the performance of membrane increase. Besides, the addition of PVP additives also enhances the membrane performance for certain membrane sample. However, there is other parameter that should be taking into account such as the thickness of membrane, the surface membrane chemistry, size of organic matter, fouling etc. which could affect the performance of membrane.
3. The percentage of COD removal in this study show low rejection while the percentage of TDS removal shows high rejection.
4. The SEM images show the structural illustration of membrane n term of cross section and top surface. By analyzing this data, it helps in improving the understanding on how the membranes perform.

Acknowledgement

This work is supported via financial fund namely the Fundamental Research Grants Scheme (FRGS): 600-RMI/FRGS 5/3 (75/2015) managed by Research Management Centre, Universiti Teknologi MARA, Shah Alam, Malaysia.

References

1. N. Descoins, S. Deleris, R. Lestienne, E. Trouvé, and F. Maréchal, "Energy efficiency in waste water treatments plants: Optimization of activated sludge process coupled with anaerobic digestion," Energy, **vol. 41**, no. 1, pp. 153–164, 2012.
2. B. H. Diya'Uddeen, W. M. A. W. Daud, and a. R. Abdul Aziz, "Treatment technologies for petroleum refinery effluents: A review," *Process* Saf. Environ. Prot., **vol. 89**, no. 2, pp. 95–105, 2011.
3. C. L. Yang, "Electrochemical coagulation for oily water demulsification," Sep. Purif. Technol., **vol. 54**, no. 3, pp. 388–395, 2007.
4. R. Davarnejad, M. Mohammadi, and A. F. Ismail, "Petrochemical wastewater treatment by electro-Fenton process using aluminum and iron electrodes: Statistical comparison," J. Water Process Eng., **vol. 3**, pp. 18–25, 2014.
5. M. M. Rahman and M. H. Al-Malack, "Performance of a crossflow membrane bioreactor (CF-MBR) when treating refinery wastewater," Desalination, **vol. 191**, no. 1–3, pp. 16–26, 2006.
6. F. Shahrezaei, Y. Mansouri, A. A. L. Zinatizadeh, and A. Akhbari, "Process modeling and kinetic evaluation of petroleum refinery wastewater treatment in a photocatalytic reactor using TiO 2 nanoparticles," Powder Technol., **vol. 221**, pp. 203–212, 2012.
7. C. Chen, L. Wei, X. Guo, S. Guo, and G. Yan, "Investigation of heavy oil refinery wastewater treatment by integrated ozone and activated carbon - supported manganese oxides," Fuel Process. Technol., **vol. 124**, pp. 165–173, 2014.
8. S. H. Lin and W. J. Lan, "Waste oil/water emulsion treatment by membrane processes," J. Hazard. Mater., **vol. 59**, no. 2–3, pp. 189–199, 1998.
9. B. Chakrabarty, a. K. Ghoshal, and M. K. Purkait, "Ultrafiltration of stable oil-in-water emulsion by polysulfone membrane," J. Memb. Sci., **vol. 325**, pp. 427–437, 2008.
10. E. Saljoughi and S. M. Mousavi, "Preparation and characterization of novel polysulfone nanofiltration membranes for removal of cadmium from contaminated water," *Sep. Purif. Technol.*, **vol. 90**, pp. 22–30, 2012.
11. W. L. Ang, A. W. Mohammad, N. Hilal, and C. P. Leo, "A review on the applicability of integrated/hybrid membrane processes in water treatment and desalination plants," Desalination, 2014.
12. A. Salahi, A. Gheshlaghi, T. Mohammadi, and S. S. Madaeni, "Experimental performance evaluation of polymeric membranes for treatment of an industrial

oily wastewater," Desalination, **vol. 262**, no. 1–3, pp. 235–242, 2010.

13. A. Salahi, T. Mohammadi, and R. Mosayebi, "Journal of Environmental Chemical Engineering Asymmetric polyethersulfone ultra fi ltration membranes for oily wastewater treatment : Synthesis , characterization , ANFIS modeling , and performance," Biochem. Pharmacol., **vol. 3**, no. 1, pp. 170–178, 2015.
14. A. Pagidi, R. Saranya, G. Arthanareeswaran, a. F. Ismail, and T. Matsuura, "Enhanced oil-water separation using polysulfone membranes modified with polymeric additives," Desalination, **vol. 344**, pp. 280–288, 2014.
15. A. F. Ismail and A. R. Hassan, "Effect of additive contents on the performances and structural properties of asymmetric polyethersulfone (PES) nanofiltration membranes," Sep. Purif. Technol., **vol. 55**, no. 1, pp. 98–109, 2007.
16. C. Barth, M. C. Gonçalves, a. T. N. Pires, J. Roeder, and B. a. Wolf, "Asymmetric polysulfone and polyethersulfone membranes: Effects of thermodynamic conditions during formation on their performance," J. Memb. Sci., **vol. 169**, no. 2, pp. 287–299, 2000.
17. Y. Zheng, S. Yu, S. Shuai, Q. Zhou, Q. Cheng, M. Liu, and C. Gao, "Color removal and COD reduction of biologically treated textile effluent through submerged filtration using hollow fiber nanofiltration membrane," Desalination, **vol. 314**, pp. 89–95, 2013.
18. B. Chakrabarty, a. K. Ghoshal, and M. K. Purkait, "Preparation, characterization and performance studies of polysulfone membranes using PVP as an additive," J. Memb. Sci., v**ol. 315**, no. 1–2, pp. 36–47, 2008.
19. M. J. Han and S. T. Nam, "Thermodynamic and rheological variation in polysulfone solution by PVP and its effect in the preparation of phase inversion membrane," J. Memb. Sci., **vol. 202**, no. 1–2, pp. 55–61, 2002.

In-Situ Catalytic Surface Modification of Micro-Structured $La_{0.6}Sr_{0.4}Co_{0.2}Fe_{0.8}O_{3-\delta}$ (LSCF) Oxygen Permeable Membrane Using Vacuum-Assisted technique

Nur Hidayati Othman[1], Munawar Zaman Shahruddin[1], 'Ainun Sailah Sihar[1], Zhentao Wu[2] and K. Li[2]

[1]*Department of Oil and Gas Engineering, Faculty of Chemical Engineering, Universiti Teknologi MARA, 4050 Shah Alam, Malaysia*
[2]*Department of Chemical Engineering, Imperial College London, London SW7 2AZ, UK*

Abstract. This paper aims at investigating the means to carry out in-situ surface modification of $La_{0.6}Sr_{0.4}Co_{0.2}Fe_{0.8}O_{3-\delta}$ (LSCF) oxygen permeable membrane by using vacuum assisted technique. The unique structure of the LSCF hollow fibre membrane used in this study, which consists of an outer dense oxygen separation layer and conical-shaped microchannels open at the inner surface has allowed the membrane to be used as oxygen separation membrane and as a structured substrate for where catalyst can be deposited. A catalyst solution of similar material, LSCF was prepared using sol-gel technique. Effects of calcination temperature and heating rate were investigated using XRD and TGA to ensure pure perovskites structure of LSCF was obtained. It was found that a lower calcination temperature can be used to obtain pure perovskite phase if slower heating rate is used. The SEM photograph shows that the distribution of catalyst onto the membrane microchannels using in-situ deposition technique was strongly related to the viscosity of LSCF catalytic sol. Interestingly, it was found that the amount of catalyst deposited using viscous solution was slightly higher than the less viscous sol. This might be due to the difficulty of catalyst sol to infiltrate the membrane and as a result, thicker catalyst layer was observed at the lumen rather than onto the conical-shaped microchannels. Therefore, the viscosity of catalyst solution and calcination process should be precisely controlled to ensure homogeneous catalyst layer deposition. Analysis of the elemental composition will be studied in the future using energy dispersive X-ray Spectroscopy (EDX) to determine the elements deposited onto the membranes. Once the elemental analysis is confirmed, oxygen permeation analysis will be carried out.

1 Introduction

Currently, industrial oxygen production relies on traditional technologies such as cryogenic distillation and pressure swing adsorption, which required a prohibitively large capital investment and high operating costs. Therefore, a potentially cheaper technology by using dense mixed ionic and electronic conducting (MIEC) ceramic membrane to separate oxygen from air is needed [1, 2]. Through oxygen enrichment of the combustion air with ceramic membranes, significant saving of energy can be expected due to enhanced combustion efficiency.

MIEC oxides are multifunctional materials that are widely used as catalysts [3], fuel cells electrodes [4], and oxygen permeation membranes [5]-[7]. In compared to the oxygen permeation over polymer-based membrane and industrial state-of-the-art method, the membranes developed using MIEC oxides have significant advantages such as simplified operation and high efficiency. Besides that, this MIEC dense membrane has 100% oxygen permeation selectivity at elevated temperatures. They allow only oxygen to diffuse through vacancies in the crystal lattice, which is charge-compensated by intrinsic electronic conductivity. Thus, various applications behold such as for oxygen production and as well as many industrial processes that require constant supply of oxygen i.e, conversion of hydrocarbon (oxidative coupling methane [8], [9] and partial oxidation of methane to syngas [10]) and oxygen ion conductor fuel cell. Besides that, dense MIEC ceramic membranes exhibit both excellent thermal stability and high oxygen flux which is thus make them suitable in the field of energy production or in integrated coal gasification power station [11], [12].

The fundamental studies of these MIEC membranes started with the pioneering work from Teraoka and co-workers in 1980's [13]. Perovskite type membranes having MIEC have been considered as one of the attractive materials for the oxygen separation. They could produce pure oxygen without the need for electrodes and external electrical loadings. Among various types of perovskites, $La_{1-x}Sr_xCo_{1-y}Fe_yO_{3-\delta}$ (LSCF) has been considered as one of the promising materials due to its high oxygen permeation flux and stability. However, improvement of oxygen permeation flux is still needed for future commercialization.

It is known that the overall oxygen transport through a dense MIEC membrane are controlled by two mechanisms; (1) surface exchange reactions at the membrane surfaces and (2) oxygen diffusion rate through the bulk membrane (Fig. 1). Thus, the oxygen diffusion rate across the membrane can be improved by using better MIEC materials and/or reducing the membrane thickness. Asymmetric membrane structures, where a thin layer of dense MIEC deposited onto porous substrates have been widely proposed in order to reduce the effect of bulk diffusion. However, as the membrane thickness decreases, the effects of surface exchange reaction become dominating. Several studies have investigated the effects of adding porous catalytic layers on either side or both sides of the membrane and most of them resulting higher oxygen flux. This is due to the modification of oxygen potential drops at gas-membrane interfaces and improvement of oxygen exchange rates by the catalytic layers [14]-[18].

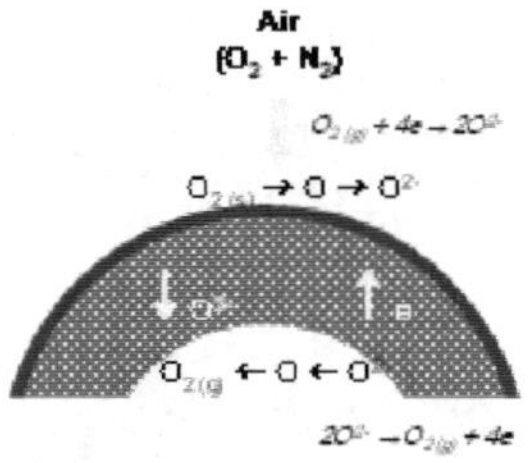

Figure 1. Mechanism of oxygen transport through dense MIEC

It has been realised that not all MIEC membranes are suitable to be fabricated into thin self-supported hollow fibre configuration due to mechanical strength and chemical stability. Among the perovskites materials under thoroughly studies are $La_{0.6}Sr_{0.4}Co_{0.2}Fe_{0.8}O_{3-\delta}$ (LSCF)[19, 20] and $Ba_{0.5}Sr_{0.6}Co_{0.2}Fe_{0.8}O_{3-\delta}$ (BSCF)[7, 21], which are very promising as oxygen separation membrane. In our previous works [9, 22], it was shown that a unique asymmetric structure of LSCF hollow fibre membrane, which consists of conical-shaped microchannels open at the inner surface and an outer dense oxygen separation can be prepared via a viscous fingering induced phase inversion. Not only this design has substantially reduced resistance across the membrane, but the open microchannels can act as a structured substrate where catalyst can be deposited/coated for enhancement of surface activity. As the surface area to volume ratio of this hollow fibre configuration is extremely high in compared to other configurations (disc & tubular), the performance of this membrane can greatly exceed that of other membrane systems. The thin hollow fibres also show a better mechanical stability in compared to disk-shaped membranes of the same wall thickness.

Fig. 2 shows a hollow fibre membrane with typical symmetric and asymmetric structures. The catalyst can be easily coated on the outer surface of the hollow membrane by using dip coating [23] or brushing techniques[24]. However, for the micro-structured LSCF hollow fibre membrane with open microchannels, the techniques stated earlier were unlikely to provide uniform catalyst coating inside the lumen of the hollow fibre membrane. Therefore, a new vacuum-assisted technique for in-situ catalyst deposition has been proposed in the present work.

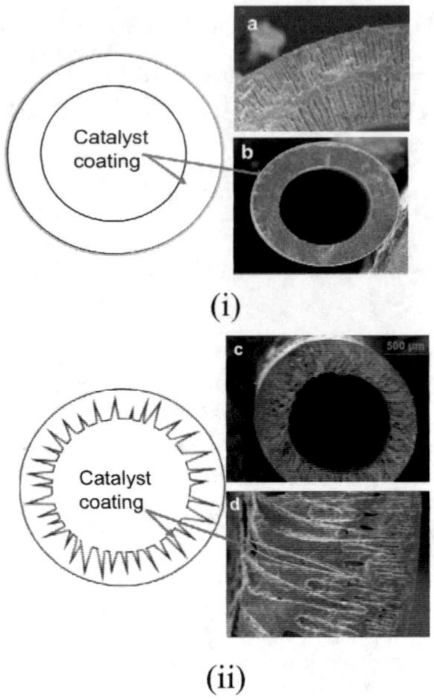

Figure 2. Surface modification of oxygen permeable membrane in (i) conventional a) asymmetric and b) symmetric structure and (ii) open microchannels structure (micro-structured).

2 Methodology

This work was divided into two main parts, which are preparation of the catalyst using sol-gel method and in-situ deposition of catalyst onto the micro-structured LSCF membrane microchannels using vacuum assisted techniques. The LSCF hollow fibre membrane used in this work was fabricated using a viscous fingering induced phase inversion process as described by Othman et al. Several characterization techniques such were then used to analyse the synthesized catalyst and surface-modified membrane.

2.1 Preparation of LSCF perovskites catalyst via sol-gel technique

The LSCF catalyst was prepared using the corresponding metal nitrates: $La(NO_3)_3.6H_2O$ (EMD Millipore) $Sr(NO_3)_{2v}$ (EMD Millipore,), $Co(NO_3)_2.6H_2O$ (Systerm ChemAR), and $Fe(NO_3)_3.9H2O$ (R&M Chemical). The starting solution was prepared by dissolving appropriate amounts of metal nitrate into distilled water by stirring at 80 °C. Citric acid and ethylene glycol were then added to the fully dissolved nitrate solution at the mole ratio of total metal ions: citric acid: ethylene glycol= 9: 3: 1. The purpose of adding ethylene glycol is to inhibit cation segregation, resulting a homogeneous precursor solution. Then, the resultant solution was continuously stirred at 80 °C to accelerate polyesterification reaction of citric acid and ethylene glycol and continuous polymerization until a transparent brownish LSCF catalytic sol was formed. The LSCF catalytic sol was dried at 45 °C for 4 hours and 24 hours to investigate the effects of catalytic sol-viscosity. The LSCF sol was then calcinated at various calcination temperatures (600 – 900 °C) and heating rates (5.0 °C/min and 2.0 °C/min). Fig. 3 shows the scheme of LSCF preparation.

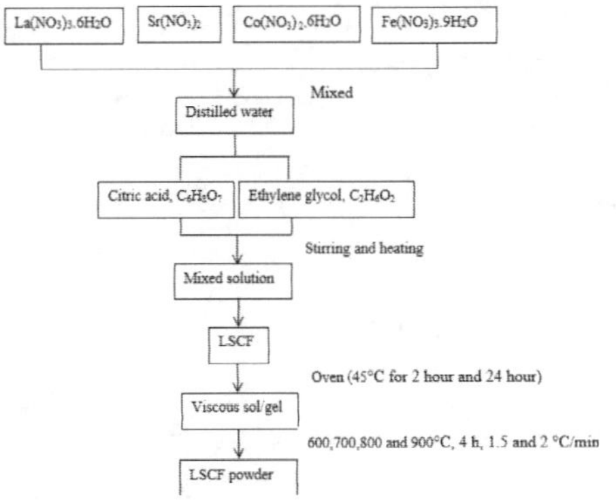

Figure 3. Scheme of preparation of LSCF using sol-gel technique

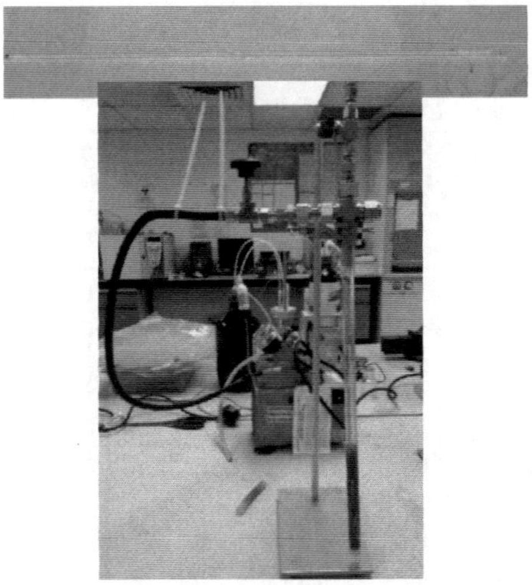

Figure 4. Photograph on thee in-situ deposition of LSCF using vacuum-assisted technique

2.2 In-situ deposition of catalyst onto micro-structured LSCF oxygen permeable membrane

Prior to the deposition of catalysts, the micro-structured LSCF hollow fibre membranes thoroughly cleaned with water and acetone and dried in air. Then, the membranes were wrapped with PTFE tape to avoid any contamination on the outer surface of the hollow fibre during the coating. A vacuum assisted technique was used to improve the uniformity and controlled the deposition of catalyst inside the lumen of LSCF hollow fibre membrane. The main function of the vacuum pump was to aid the removal of air inside the hollow fibre. Fig. 4 shows the in-house vacuum-depositor system used in this work. After the deposition, the fibres were dried at room temperature for 1 hour prior to calcination process.

2.3 Characterizations

The phase development of LSCF at different calcination temperatures and heating rates were analysed using X-ray diffractometer (XRD). The XRD patterns were obtained with an X'celerator detector (Rigaku) model using Cu-Kα as the radiation source. The XRD scans were carried out in 2θ which range from 10° to 80° using a step width of 0.05°. Thermogravimetric analysis of the synthesized catalysts was carried out by using a TG analyzer (Mettler Toledo) from room temperature to 1000 °C in air at a heating/cooling rate of 10 °C/min. The composition and purity of the synthesized. Both morphology and structure of the LSCF micro-structured hollow fibre membranes as well as the deposition of LSCF catalyst were observed by scanning electron microscopy (SEM) on HITACHI S-4800 (Japan).

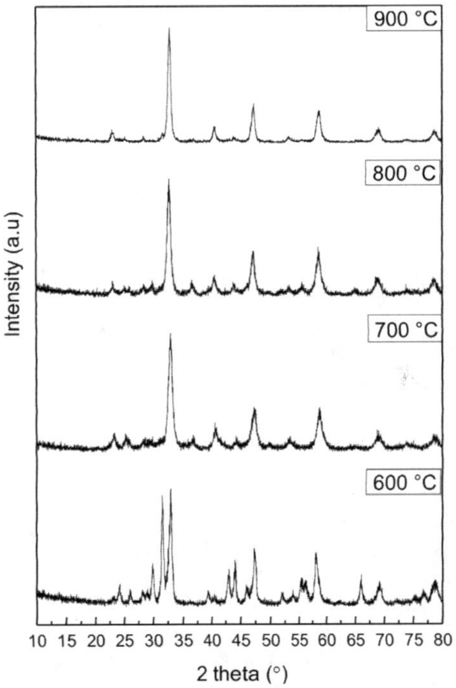

Figure 5. XRD powder patterns of LSCF powder calcined at 600-900°C

3 Results and discussions

3.1 Catalyst synthesis

3.1.1 LSCF

Calcination temperature could greatly affect the catalyst activity. A higher calcination temperature is usually required to obtain pure crystalline perovskite structure. However, with the increase of calcination temperatures the porosity and surface area of the perovskite catalyst could be reduced. Fig. 5 shows the effect of calcination temperatures on crystalline structure of LSCF at a heating rate of 5 ° C/min. It can be seen that at higher temperature orthorhombic perovskite structure was obtained and at 900 °C, pure perovskite phase was developed by elimination of $SrCO_3$ [25]. In addition, the crystallinity of diffraction peaks was observed to increase with calcination temperatures. The higher intensity of the corresponding characteristic peaks of LSCF was due to the formation of larger crystalline particles after calcination process. The peaks of LSCF for were observed to shift slightly towards higher 2-theta values as the calcination temperatures increased. This indicates that the lattice diameter of LSCF decreases during the calcination process.

Thermal decomposition usually takes place in three stages. The first decomposition stage is at temperature around 100-250 °C due to the loss of adsorbed water. At temperature between 300 to 750 °C, second decomposition stages take place, which can be attributed to the burnout of organic components and third stages is the pyrolysis reaction may be due to complete dissociation of carbonates produced during combustion and initiation of the formation of LSCF phase. Fig. 6 shows the thermogravimetric analysis (TGA) curve of the calcined LSCF at various temperatures while Table 1 reports the mass loss values. The percentage of mass loss for the samples that were calcinated above 800 °C was found to be slightly lower (less than 4%) in compared to other samples because all the organics and impurities in the LSCF has been removed through calcination process. The small percentage of mass loss observed for temperature above 800 °C was believed as a result of $SrCO_3$ decomposition.

3.1.2 Effects of heating rate

In order to investigate the effect of heating rate, the LSCF sol were calcined at 800 °C using two different heating rates, which were 2 °C/min and 5 °C/min. Fig. 7 shows the XRD powder pattern of the calcinated LSCF powders. From the XRD pattern obtained, it is clearly shown that the formation of a single perovskite phase was affected by the heating rates used during the calcination process. At slower heating rate (2 °C/min), a clear crystalline perovskite structure can be formed at much lower temperature in compared to a faster heating rate (5 °C/min). This was because the sample has longer time to remove organic binder and impurities during heating up period [26]. Fig. 8 compares the amount of time required to complete the calcination process at 800 °C. The total calcination time when heating rate of 2 °C/min was used is 1010 minutes, while 548 minutes is required if a heating rate of 5 °C/min was used. This indicates that the shorter calcination time was not sufficient in removing the impurities, thus lead to an incomplete perovskite structure. Therefore, by using slower heating rate, lower calcination temperature can be used.

Table 1 TGA weight loss percentage of LSCF

Calcination temperatures (°C)	Weight loss (%)
600	8.0
700	5.1
800	4.2
900	0.1

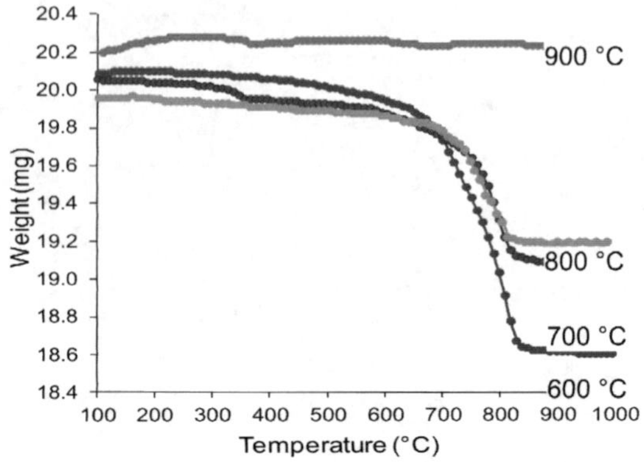

Figure 6. TGA plot of LSCF calcined at 600-900 ° C

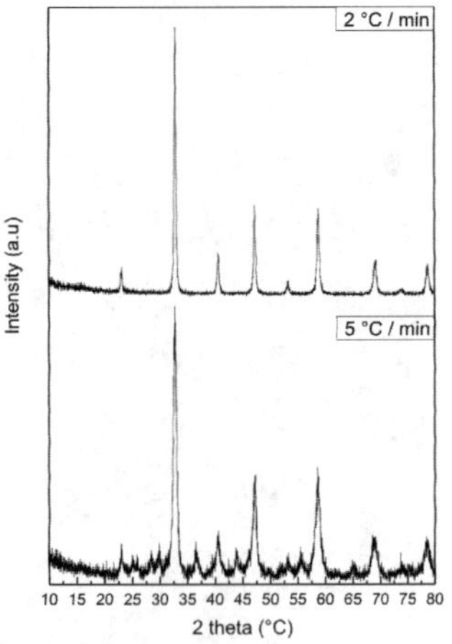

Figure 7. XRD powder patterns of LSCF powder sintered at 800 °C with 2 and 5 °C/min heating rate.

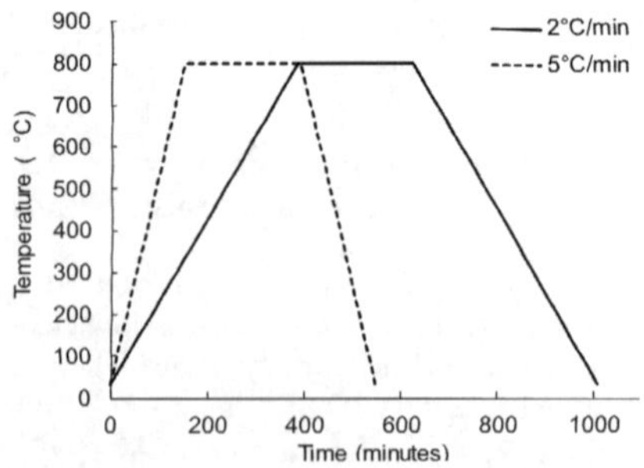

Figure 8. Total time required for the calcination process using heating rate of 2 °C/min and using 5 °C/min.

DOI: 10.1051/matecconf/20166905002

3.2 In-situ deposition of LSCF catalyst inside the micro-structured LSCF membrane

Viscosity of catalyst solutions plays a very role in order to obtain thin and homogeneous catalyst layer [27, 28]. If the catalyst sol is too low, the catalyst could not be deposited onto the membrane surface (dense membrane) or in a case of porous membrane, the catalyst sol will easily infiltrate into the membrane pores. Conversely, the use of highly viscous sol could lead to non-homogeneous deposition of catalyst that could lead to cracks of catalyst layer. The morphology and homogeneity of in-situ deposited LSCF catalyst was studied by SEM. Three different LSCF sol were used in this work. The viscosity of the sol was varied by increasing the evaporation time. The first sol (S1) used was without any evaporation, while the second sol (S2) and third sol (S3) was dried in an oven for 4 hours and 24 hours, respectively. After the deposition, LSCF membrane was calcined. Fig. 9 shows the SEM images of S1. A good coverage of catalyst onto the open conical-shaped microchannels was observed. However, if the depositing process is not properly controlled, there is a tendency of catalyst layer to crack and delaminated especially at the top of conical-shaped (Fig. 9a) and entrance (Fig. 9b) of microchannels.

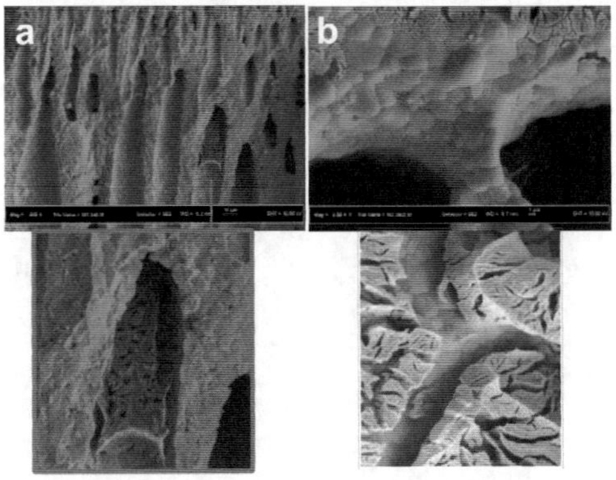

Figure 9. SEM photograph showing the surface morphology of S1 with a) top of conical-shaped microchannels and b) entrance of the microchannels.

Fig. 10 compares the SEM images of the in-situ deposited LSCF onto the micro-structured LSCF oxygen permeable membrane using the vacuum assisted technique. It can be seen clearly that the LSCF catalyst was successfully deposited without blocking the entrance to microchannels LSCF membrane. However, as the viscosity of the catalyst sol is higher (S3>S2>S1), uneven catalyst distribution across the membrane lumen can be seen. Interestingly, it was found that the amount of catalyst deposited using viscous solution was slightly higher than the less viscous sol. This might be due to the difficulty of catalyst sol to infiltrate the membrane and as a result, thicker catalyst layer was observed at the microchannels entrance rather than onto the conical-shaped microchannels (Fig. 10-S3). Therefore, the viscosity of catalyst solution and calcination process should be precisely controlled in the future to ensure homogeneous catalyst layer deposition.

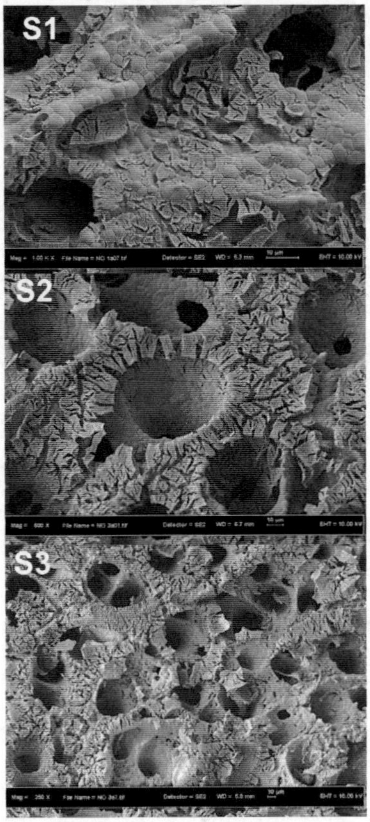

Figure 10. SEM photograph of in-situ deposited LSCF catalyst using three different catalytic sol evaporation time. S1) No evaporation, S2) 4 hours of evaporation and S3) 24 hours of evaporation

4 Conclusions

In this work, LSCF catalytic sol was first studied to investigate the effects of calcination temperature and heating rates towards the structure of LSCF. A pure perovskite structure is required to ensure high oxygen permeability. The LSCF sol was then deposited onto the unique asymmetric LSCF hollow fibre membrane, which consists of conical-shaped microchannels open at the inner surface and an outer dense oxygen separation. Due to the difficulty in obtaining thin and homogeneous catalyst layer by using conventional dip coating or brushing technique, a new vacuum-assisted technique for in-situ catalyst deposition was proposed. This preliminary study has confirmed the possibility of using the vacuum-assisted technique as one of the ways to deposited the catalyst. More details characterization such as analysis of elemental composition will be carried out in the future using energy dispersive X-ray Spectroscopy (EDX) to investigate the element deposited onto the membranes. Once the elemental analysis is confirmed, oxygen permeation analysis will be carried out.

Acknowledgement

The authors would like to thank Ministry of Higher Education Malaysia and Universiti Teknologi MARA (UiTM) for the financial support given through Research Acculturation Grant Scheme (RAGS) 600-RMI/RAGS 5/3 (54/2015)

References

1. J. Sunarso, et al., "Mixed ionic–electronic conducting (MIEC) ceramic-based membranes for oxygen separation," Journal of membrane science, vol. 320, pp. 13-41, 2008.
2. X. Tan, et al., "Mixed conducting ceramic hollow-fiber membranes for air separation," AIChE Journal, vol. 51, pp. 1991-2000, 2005.
3. Z. Q. Yu, et al., "Oxidative Coupling of Methane over Na2WO4/CeO2 and Related Catalysts," Journal of Catalysis, vol. 154, pp. 163-173, 1995.
4. T. Li, et al., "Single-step fabrication and characterisations of triple-layer ceramic hollow fibres for micro-tubular solid oxide fuel cells (SOFCs)," Journal of Membrane Science, vol. 449, pp. 1-8, 2014.
5. S. Engels, et al., "Oxygen permeation and stability investigations on MIEC membrane materials under operating conditions for power plant processes," Journal of membrane science, vol. 370, pp. 58-69, 2011.
6. G. Zhang, et al., "A novel Nb2O5-doped SrCo0.8Fe0.2O3−δ oxide with high permeability and stability for oxygen separation," Journal of membrane science, vol. 405–406, pp. 300-309, 2012.
7. Q. Jiang, et al., "Oxygen permeation study and improvement of Ba0.5Sr0.5Co0.8Fe0.2Ox perovskite ceramic membranes," Journal of membrane science, vol. 369, pp. 174-181, 2011.
8. O. Czuprat, et al., "Oxidative Coupling of Methane in a BCFZ Perovskite Hollow Fiber Membrane Reactor," Industrial & engineering chemistry research, vol. 49, pp. 10230-10236, Nov 3 2010.
9. N. H. Othman, et al., "A micro-structured La 0.6 Sr 0.4 Co 0.2 Fe 0.8 O 3− δ hollow fibre membrane reactor for oxidative coupling of methane," Journal of Membrane Science, vol. 468, pp. 31-41, 2014.
10. N. H. Othman, et al., "Bi1.5Y0.3Sm0.2O3-δ-based ceramic hollow fibre membranes for oxygenseparation and chemicalreactions," Journal of Membrane Science, vol. 432, pp. 58-65, 2013.
11. J. Tonziello and M. Vellini, "Oxygen production technologies for IGCC power plants with CO2 capture," Energy Procedia, vol. 4, pp. 637-644, 2011.
12. B. Jin, et al., "Plantwide control and operating strategy for air separation unit in oxy-combustion power plants," Energy Conversion and Management, vol. 106, pp. 782-792, 2015.
13. Y. Teraoka, "Mixed ionic-electronic conductivity of La1-xSrxCo1-yFeyO3-[delta] perovskite-type oxides," Materials research bulletin, vol. 23, p. 51, 1988.
14. Z. Wang, et al., "Improvement of the oxygen permeation through perovskite hollow fibre membranes by surface acid-modification," Journal of Membrane Science, vol. 345, pp. 65-73, 2009.
15. A. Leo, et al., "High performance perovskite hollow fibres for oxygen separation," Journal of membrane science, vol. 368, pp. 64-68, 2011.
16. V. V. Kharton, et al., "Surface modification of La0.3Sr0.7CoO3−δ ceramic membranes," Journal of Membrane Science, vol. 195, pp. 277-287, 2002.
17. A. Leo, et al., "Oxygen permeation through perovskite membranes and the improvement of oxygen flux by surface modification," Science and Technology of Advanced Materials, vol. 7, pp. 819-825, 2006.
18. L. Olivier, et al., "Oxidative coupling of methane using catalyst modified dense perovskite membrane reactors," Catalysis today, vol. 142, pp. 34-41, 2009.
19. X. Tan, et al., "Morphology control of the perovskite hollow fibre membranes for oxygen separation using different bore fluids," Journal of membrane science, vol. 378, pp. 308-318, 2011.
20. X. Tan, et al., "Oxygen production using La0.6Sr0.4Co0.2Fe0.8O3−α (LSCF) perovskite hollow fibre membrane modules," Journal of membrane science, vol. 310, pp. 550-556, 2008.
21. A. Leo, et al., "The enhancement of oxygen flux on Ba 0.5 Sr 0.5 Co 0.8 Fe 0.2 O 3− δ (BSCF) hollow fibers using silver surface modification," Journal of Membrane Science, vol. 340, pp. 148-153, 2009.
22. N. H. Othman, et al., "An oxygen permeable membrane microreactor with an in-situ deposited Bi 1.5 Y 0.3 Sm 0.2 O 3− δ catalyst for oxidative coupling of methane," Journal of Membrane Science, vol. 488, pp. 182-193, 2015.
23. H. Tikkanen, et al., "Examination of the co-sintering process of thin 8YSZ films obtained by dip-coating on in-house produced NiO-YSZ," Journal of the European Ceramic Society, vol. 31, pp. 1733-1739, Aug 2011.
24. V. Meille, "Review on methods to deposit catalysts on structured surfaces," Applied Catalysis A: General, vol. 315, pp. 1-17, 2006.
25. S. Liu, et al., "Synthesis and characterization of La0. 8Sr0. 2Co0. 5Fe0. 5O3±δ nanopowders by microwave assisted sol–gel route," Journal of Sol-Gel Science and Technology, vol. 44, pp. 187-193, 2007.
26. P. Bomlai, et al., "Effect of heating rate on the properties of Sb and Mn-doped barium strontium titanate PTCR ceramics," Materials Letters, vol. 59, pp. 118-122, 2005.
27. J. Bravo, et al., "Wall coating of a CuO/ZnO/Al 2 O 3 methanol steam reforming catalyst for micro-channel reformers," Chemical Engineering Journal, vol. 101, pp. 113-121, 2004.
28. R. Sonawane, et al., "Preparation and photo-catalytic activity of Fe TiO 2 thin films prepared by sol - gel dip coating," Materials Chemistry and Physics, vol. 85, pp. 52-57, 2004.

MATEC Web of Conferences **69**, 05003 (2016) DOI: 10.1051/matecconf/20166905003
ICCPE 2016

Photocatalytic Degradation of Oil using Polyvinylidene Fluoride/Titanium Dioxide Composite Membrane for Oily Wastewater Treatment

Ummi Nadiah Rusli, Nur Hashimah Alias, Munawar Zaman Shahruddin, Nur Hidayati Othman

Faculty of Chemical Engineering, Universiti Teknologi Mara, 40500 Shah Alam, Selangor, Malaysia

Abstract. Production of industrial wastewater is increasing as the oil and gas industry grows rapidly over the years. The constituents in the industrial wastewater such as organic and inorganic matters, dispersed and lubricant oil and metals which have high toxicity become the major concern to the environment and ecosystem. There are many technologies are being used for oil removal from industrial wastewater. However, there are still needs to find an effective technology to treat oily wastewater before in can be discharge safely to the environment. Membrane technology is an attractive separation technology to treat oily wastewater. The aim of this study is to fabricate polyvinylidene/titanium dioxide (PVDF/TiO_2) composite membrane with further treatment using hot pressed method to enhance the adhesion between TiO_2 with the membrane surfaces. In this study the structural and physical properties of fabricated membrane were conducted using X-ray diffraction (XRD) and Fourier transform infrared spectroscopy (FTIR) respectively. The photocatalytic degradation of oil was measured using UV-Vis Spectroscopy. The FTIR results confirmed that, hot pressed PVDF/TiO_2 membrane TiO2 was successfully deposited onto PVDF membranes surface and XRD results shows that the XRD pattern of PVDF//TiO_2 found that the crystalline structure was remained unchanged after hot pressed. Clear water was obtained after synthetic oily wastewater was exposed to visible light for at least 6 hours. In conclusion, PVDF/TiO_2 composite membrane can be a potential candidate to degrade oil in oily wastewater and suggested to possess an excellent performance if perform simultaneously with membrane separation process.

1 Introduction

Oil and gas extraction operations produced wastewater which is constitutes the industry's most significant waste stream on the basis of volume. Annually, the oil and gas industry produces which is about 14 billion bbls of water (Siriverdin and Dallbauman, 2008). Wastewater contains many constituents such as organic and inorganic matters, dispersed oil, lubricant oil, chemicals and metals which may be toxic to environment. Discharging industrial wastewater without being treated can contaminate water and soil which will lead disturbances to the ecosystem. Membrane technology is an attractive separation technology (X. Zhang, Wang, & Diniz Da Costa, 2014).

The advantage of this technology is it has a high energy efficiency. Membrane system has a high area-to-volume ratio and it is easy to scale-up because of its modular design (Rajovic et al., 2015). The membrane properties, such as pore size distribution and porosity will also affect the membrane performance as an absorption and separation technology. Thus, good pore size and narrow pore size distribution of microporous membranes are required (X. Zhang et al., 2014) to ensure the high permeate flux and rejection in membrane separation process. Recently, the most popular and available membrane materials are polytetrafluoroethylene (PTFE), polypropylene (PP), polyethersulfone (PES) and polyvinylidene fluoride (PVDF). In membrane separation fields, porous membranes are commonly applied in membrane distillation, supported liquid membrane, osmotic distillation, membrane based gas absorption, membrane contact and others. Polyvinylidene fluoride (PVDF) is one of the most commonly used as polymer membrane due its excellent properties such as chemical resistance, thermal stability and good mechanical strength (Lee, Elam, & Darling, 2015). PVDF is a semi-crystalline polymer where the crystalline phase offers mechanical strength and impact resistance and the amorphous phase provides flexibility (Liu, 2011).

Porous PVDF membranes can be produced through a phase inversion method (Damodar, You, & Chou, 2009), where it is the common industrial processes in large scale membrane production. Furthermore, PVDF is soluble in high-boiling point and organic solvents such as N-methyl-2-pyrrolidone (NMP), N,N-dimethylacetamide (DMAC) and N,Ndimethylformamide (DMF), which make it easier to produce more permeable asymmetric membranes through the dry/wet phase inversion process. Besides, the high mechanical strength of PVDF (Yu, Shen, & Xu, 2009) makes PVDF membranes suitable in water application and wastewater treatment process (Ji et al., 2015).

Titanium dioxide (TiO_2) is a metal oxide semiconductor used in environment applications due to its non-toxic, strong ultraviolet (UV) absorption, good stability in photocatalyst reaction and commonly energy band gap of 3.2 eV (S. Zhang et al., 2013 and Suphankij et al., 2013). Photocatalysis with semiconductor materials has been widely explored among others advanced oxidation processes to treat toxic organic compounds in wastewater (Berry & Mueller, 1994). In specific, photocatalysis based on titanium dioxide (TiO_2) has involved a great deal of consideration due to its functional properties such as chemical and thermal stability and strong mechanical properties (Yu et al., 2009). In order to enhance the adhesion between the TiO_2 with membrane, and to further increase the efficiency of the membrane, pre-treatments such as hot press have been used (Kim et al., 2013). Recently, mechanical pressure was applied to break outer sheaths, thereby exposing the membrane accomplishing a high surface area (Liu, Chu, & Ding, 2004). As the oil and gas industry in which the refinery process or any petrochemical industries generate a big amount of wastewater, the wastewater treatment is significant in order to meet environmental standard set by government. A proper treatment of wastewater is a must as it will affect the environment. Thus, this paper presents a preliminary study on the performances of TiO_2 composite PVDF membrane with hot press treatment on photocatalytic degradation of oil in oily wastewater.

2 Methodology

2.1. Materials

The material used in this research were commercial polyvinylidene fluoride (PVDF) pellets (Kynar® 740) purchased from Arkema Inc. USA as the polymer for membrane formation, N-N-dimethylacetamide (DMAC) used as the solvent and distilled water used as coagulants during membrane fabrication. TiO_2 powder (Sigma-Aldrich) was used as additive in membrane dope.

2.2. Membrane Preparation

Dope solution was prepared by dissolving PVDF (13 wt%) pellet with DMAC (85 wt%) in a scott bottle and stirred at 60 ℃ and 600 rpm overnight in order to produce homogeneous solution. Then, TiO_2 (2 wt%) powder was added into dope solution and stirred for 2 hours. The prepared dope solution was sonicated for 1hour to remove micro bubbles. After that, prepared dope solution was casted on a glass plate with the thickness of 300 μm. After 30 second exposed to the air, the plate was immersed into deionized water coagulation bath and the membrane was slowly formed due to polymer precipitation. After 10 minutes or when the membrane was started to detached from the plate, the membrane was soaked into deionized water overnight to remove the excessive solvent before it can be dried and ready to be used. The prepared membrane was further treated using hot press method to enhance the adhesion between the TiO2 with the membrane. PVDF/TiO_2 membrane was hot pressed using hot press machine at temperature of 150 °C with pressure of 30 psi for 15 minutes to form a network of large pores.

2.3. Characterization

Characterizations of the prepared membranes were carried out using x-ray diffraction (XRD) and fourier transform infrared spectroscopy (FTIR) to measure structural and physical properties of the membranes.

2.4. Photocatalytic Degradation Measurement

Synthetic oily wastewater was formulated using 90 ml of distilled water, 5 ml of lubricant oil and 5 ml of surfactant (SPAN 80) and mixed thoroughly in 200 ml beaker before sonicated for 1 hour. The membrane sample (15 cm^2) was immersed into synthetic oily wastewater under 18W fluorescent light for 4 hours. The suspensions of 3ml were collected at 60 minutes interval for analysis. UV-Vis spectroscopy was used to measure the concentration change of oil throughout experiment at 377 nm wavelength. The photocatalytic activity was measured in percentages of oil degradation using the following equation (Mohamed, 2015):

$$\text{Degradation of oil} = (C_o - C_t)/C_t \times 100 \quad (1)$$

Where C_o is and C_t are the initial concentration at time t=0 and the concentration at time (60, 120, 180, 240, 300 and 360 minutes), respectively.

3 Results and Discussion

3.1. Characterization

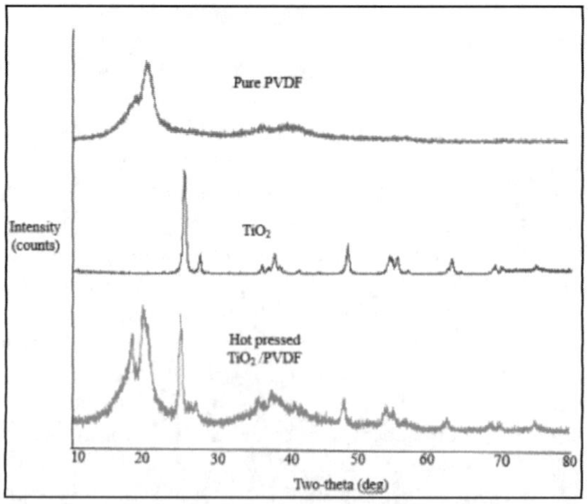

Figure 1: XRD pattern of pure PVDF and hot pressed PVDF/TiO_2 composite membrane.

DOI: 10.1051/matecconf/20166905003

The XRD patterns of PVDF/TiO_2 (1 wt%) shown in Figure 1 illustrates that the peak positions of PVDF in hot pressed PVDF/TiO_2 composite membrane exist at 2θ = 18.74 which corresponding to (0 2 0) plane and 2θ = 20.463 corresponding to (1 1 0) plane of monoclinic structure. Meanwhile, the peak positions of TiO_2 were observed at 2θ = 25.14 (1 0 1), 27.13 (1 0 1), 27.416 (1 0 1), 35.821 (1 1 2), 37.696 (1 1 2), 337.859 (1 1 2), 47.957 (2 0 0), 54.898 (1 0 5), 56.528 (1 0 5), 62.535 (2 0 4), 70.104 (2 2 0) and 75.043 (2 1 5) which identified as anatase phase. Result for hot pressed PVDF/TiO_2 membrane confirmed that the TiO_2 was successfully deposited onto PVDF membranes surface. This result is in agreement with work done by Devikala et al., (2014) mentioned that pure PVDF with monoclinic structure and was found at peak positions of 2θ = 18.48(0 2 0) and 20.18 (1 1 0). The results proved that the structure of PVDF/TiO_2 membrane was preserved after hot pressed

Figure 2 shows the FTIR spectra of pure PVDF and hot pressed PVDF/TiO_2 composite membrane. The results show that the presence of PVDF was identified by the C–F stretching vibration peak centred at 1169.08 cm^{-1} in hot pressed PVDF/ TiO_2 membrane.

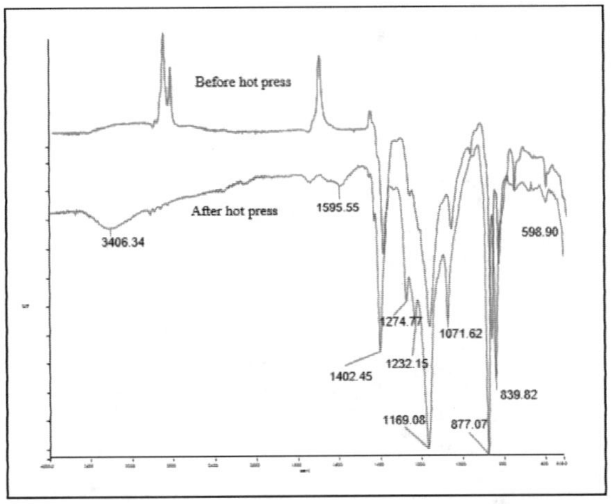

Figure 2: FTIR spectra of pure PVDF and hot pressed PVDF/ TiO_2 composite membrane

The observed vibrational band at 1402.45 cm^{-1} was related to deformation vibration of the CH_2 group and the band at 1071.62 cm^{-1} exists due to β crystalline phase of PVDF. The peaks observed at 877.07 and 839.82 cm^{-1} were rocking mode of vinylidene group of the polymer. The bands at 598.90 cm^{-1} was due to the shaking and bending vibration of CF_2. In the meantime, a band at 598.90 cm^{-1} appeared because of stretching vibration of Ti-O anatase in the composite. According to Devikala et al., (2014), the bands appeared between 800 to 1400 cm^{-1} were due to lattice vibrations of TiO_2 in the composite. The interaction between PVDF and TiO_2 make the intensity of crystalline phase of PVDF at 762 cm^{-1} peak decreased with the increase amounts of TiO_2 (Devikala et. al., 2014). The FTIR results concluded that hot press on PVDF/TiO_2 does not affect the bonding of TiO_2 on the PVDF surface as the important peak of TiO_2 remained presence at 598.90 cm^{-1}.

3.2. Photocatalytic Degradation Measurement

In photocatalytic degradation measurement oil concentrations were measured using UV-vis spectrophotometer with absorbance at 377 nm which the maximum absorption occured. During photocatalytic degradation, photo-induced electrons were produced and large amounts of OH radicals were generated because of dissociation of water molecules through oxygen vacancies. Table shows the absorbance reading for respective sampels.

Table 1: UV-Vis absorbance with respective oil concentration in photocatalytic degradation

Time	Absorbance		Oil Concentration (ppm)	
(hour)	Pure PVDF	TiO_2/PVDF	Pure PVDF	TiO_2/PVDF
0	1.377	1.377	1000	1000
1	1.252	1.184	889	839
2	1.247	1.109	885	781
3	1.271	1.122	904	791
4	1.227	0.9185	870	635
5	1.286	0.7636	915	515
6	1.263	0.5593	898	359

Figure 3 shows the effect of TiO_2 loading into PVDF membrane on oil degradation. Overall, oil concentration decreased with increased of time for both pure PVDF and PVDF/TiO_2. Oil particles were photodegraded within 6 hours of visible light from 18W fluorescent lamp source.

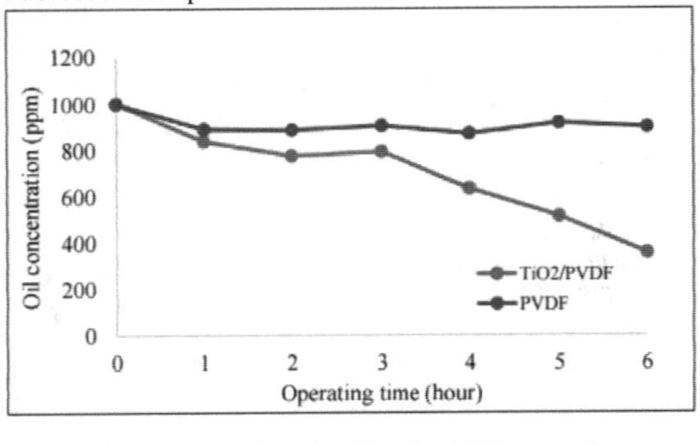

Figure 3: Effect of TiO_2 loading in PVDF membrane on oil degradation

Within 6 hours, the oil degradation using PVDF/TiO_2 composite membrane was more that 60% as compared to using soley PVDF membrane which only degraded the oil about less than 5%. This results is in agreement with result gain by Nor et al., (2016). Oil degradation using PVDF/TiO_2 composite membrane was higher as compared to pure PVDF membrane, confirmed the important role of TiO_2 catalyst in oil degradation. Owing to the photoinduced-hydrophilicity effect of TiO2 catalyst embedded within membrane matrix, the oil degradation was reported to be increased with increased of TiO_2 loading from 0 wt%, 1 wt% and 2 wt. %. Similar results have also been reported by Ong et al., 2014, where they found that membrane hydrophilicity was remarkably increased with increasing TiO_2 loading. Besides that, it was also observed that the oil degradation using pure PVDF and PVDF/TiO_2 composite membrane were increased with time. The important of oil degradation in the case of PVDF/TiO_2 composite membrane can be explained by the generation of strong oxidants from TiO_2 highest electrovalent band under visible light which oxidized most of organic compounds found in the feed solution (Ong et al., 2014). Figure 4 indicates the colour changes of oily water after 7hours exposed to visible light. From this visual observation, a clear water was obtained after at least 6hours period of time.

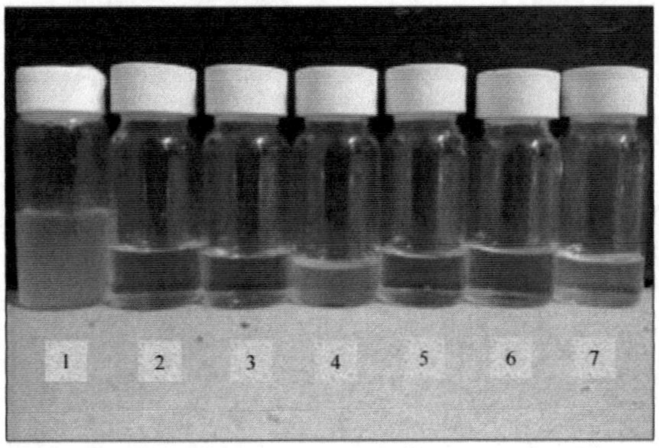

Figure 4: The colour changes of oily water after 7 hours exposed to visible light

4 Conclusions

The key to present research project was to develop a hot pressed TiO2/PVDF composite membrane and performance of degradation of oil in oily wastewater. The improvement in PVDF membrane has been achieved through the modification of the surface of PVDF membrane by incorporation with TiO2 particles membrane and further treatment using hot press method. Based on XRD diffraction peaks, the peak positions of TiO_2 after hot press appeared at 2θ = 25.14 (1 0 1), 27.13 (1 0 1), 27.416 (1 0 1), 35.821 (1 1 2), 37.696 (1 1 2), 337.859 (1 1 2), 47.957 (2 0 0), 54.898 (1 0 5), 56.528 (1 0 5), 62.535 (2 0 4), 70.104 (2 2 0) and 75.043 (2 1 5) which identified as anatase phase. Result for hot pressed PVDF/TiO_2 membrane confirmed that the TiO_2 was successfully deposited onto PVDF membranes surface. The FTIR results concluded that hot press on PVDF/TiO_2 does not affect the bonding of TiO_2 on the PVDF surface as the important peak of TiO_2 remained presence at 598.90 cm^{-1}. In photocatalytic degradation, the oil degradation using PVDF/TiO_2 composite membrane was extremely higher compared to pure PVDF membrane which suggested that the important role of TiO_2 catalyst in PVDF membrane for oil degradation. The oil degradation using PVDF/TiO_2 composite membrane was more that 60% as compared to using soley PVDF membrane which only degraded the oil about less than 5%. Clear water was obtained after synthetic oily wastewater was exposed to visible light for at least 6 hours. In conclusion, PVDF/TiO_2 composite membrane can be a potential candidates to degrade oil in oily wastewater and suggested to possess an excellent performance if perform simultaneously with membrane separation process.

Acknowledgment

The authors would like to thank the Ministry of Education (MOE) and Universiti Teknologi MARA (UiTM) for the financial support given through the Research Acculturation Grant Scheme (RAGS), 600-RMI/RAGS 5/3 (78/2013).

References

1. Berry, R. J., & Mueller, M. R. (1994). Photocatalytic Decomposition of Crude Oil Slicks Using TiO2 on a Floating Substrate. *Microchemical Journal*, *50*(1), 28–32.
2. Damodar, R. A., You, S.-J., & Chou, H.-H. (2009). Study the self cleaning, antibacterial and photocatalytic properties of TiO2 entrapped PVDF membranes. *Journal of Hazardous Materials*, *172*(2-3), 1321–8.
3. Ji, J., Liu, F., Hashim, N. A., Abed, M. M., & Li, K. (2015). Poly (vinylidene fluoride (PVDF) membranes for fluid separation. *Reactive and Functional Polymers*, *86*, 134-153.
4. Lee, A., Elam, J. W., & Darling, S. B. (2015). Membrane materials for water purification: design, development, and application. *Environ. Sci.: Water Res. Technol.*
5. Li, W. J., & Tuan, R. S. (2009). Fabrication and application of nanofibrous scaffolds in tissue engineering. *Current Protocols in Cell Biology*, 25-2.
6. Liu, F., Hashim, N. A., Liu, Y., Abed, M. M., & Li, K. (2011). Progress in the production and modification of PVDF membranes. *Journal of* Membrane *Science*, *375*(1), 1-27.
7. Liu, X., Chu, P. K., & Ding, C. (2004). Surface modification of titanium, titanium alloys, and related materials for biomedical applications. *Materials Science and Engineering R: Reports*, *47*(2004), 49–121.
8. Kim, I., Rothschild, A., Lee, B. H., Kim, D. Y., Jo, S. M., & Tuller, H. L. (2013). Ultrasensitive Chemiresistors Based on Electrospun TiO 2 Nanofibers.
9. Ong, C. S., Lau, W. J., Goh, P. S., Ng, B. C., & Ismail, A. F. (2014). Investigation of submerged membrane photocatalytic reactor (sMPR) operating parameters during oily wastewater treatment process.*Desalination*, *353*, 48-56.
10. Ong, C. S., Lau, W. J., Goh, P. S., & Ismail, A. I. (2014). Preparation and Characterization of PVDF-TiO2 Composite Membranes Blended with Different Mw of PVP for Oily Wastewater Treatment using Submerged Membrane System. *Jurnal Teknologi*, *69*(9).
11. Shi, L., Wang, R., Cao, Y., Liang, D. T., & Tay, J. H. (2008). Effect of additives on the fabrication of poly

(vinylidene fluoride-co-hexafluropropylene)(PVDF-HFP) asymmetric microporous hollow fiber membranes. *Journal of Membrane Science*, *315*(1), 195-204.

12. Shimekit, B., & Mukhtar, H. (2012). Natural gas purification technologies-major advances for CO_2 separation and future directions. *INTECH Open Access Publisher*.
13. Sun, C. (2009). *Poly (vinylidene fluoride) membranes: Preparation, modification, characterization and applications* (Doctoral dissertation, University of Waterloo).
14. Suphankij, S., Mekprasart, W., & Pecharapa, W. (2013). Photocatalytic of N-doped TiO 2 Nanofibers Prepared by Electrospinning. *Energy Procedia*, *34*, 751-756.
15. Yu, L. Yu., Shen, H. M., & Xu, Z. L. (2009). PVDF–TiO2 Composite Hollow Fiber Ultrafiltration Membranes Prepared by TiO2 Sol–Gel Method and Blending Method. *Journal of Applied Physics*, *113*(Ii), 1763–1772.
16. Zhang, S., Li, J., Zeng, M., Zhao, G., Xu, J., Hu, W., & Wang, X. (2013). In situ synthesis of water-soluble magnetic graphitic carbon nitride photocatalyst and its synergistic catalytic performance In situ synthesis of water-soluble magnetic graphitic carbon nitride photocatalyst and its synergistic catalytic performance.
17. Zhang, X., Wang, D. K., & Diniz Da Costa, J. C. (2014). Recent progresses on fabrication of photocatalytic membranes for water treatment. *Catalysis Today*, *230*, 47–54.

Pyrolysis of carbonaceous particles and properties of Carbonaceous-*g*-Poly (acrylic acid-co-acrylamide) superabsorbent polymer for agricultural applications

S. Ghazali, S. Jamari, N. Noordin and K. M. Tan

Faculty of Chemical and Natural Resources Engineering, Universiti Malaysia Pahang, 26300 Gambang Kuantan, Pahang, Malaysia

Abstract. Utilisation of fertilizer and water are very important in determining the production of agriculture nowadays. The excessive use of fertilizer in plantation somehow could leads to environmental pollution. The present study reported a synthesis of controlled release water retention (CRWR) fertilizer coating with superabsorbent polymer (SAPs). Superabsorbent polymer (SAPs) are polymers that have ability to absorb and retain large amounts of water relative to their own mass. The presence of coating layer of SAPs on fertilizer granules was believed could reduce excessive used of fertilizer by controlling their dissolution rates and also reduce the environmental pollution. In this study, the effect on the addition of carbonaceous filler in SAPs on the water absorbency was also be compared with control SAPs (without carbonaceous particles). In this study, the carbonaceous filler were obtained from pyrolysis process of empty fruit bunch (EFB) biomass. The synthesized of SAPs and carbonaceous-SAPs were carried out via solution polymerization technique by using monomer of poly(acrylic acid) (AA), acrylamide (AM), cross linker, methylene bisacrylamide (MBA) and initiator, ammonium peroxodisulfate (APS) that partially neutralized with sodium hydroxide (NaOH). The CRWR fertilizer was later be prepared by coated the fertilizer granule with SAPs and carbonaceous-SAPs. The water absorbency, morphology and the bonding formation of both CRWR fertilizer were investigated by using tea-beg method, Scanning Electron Microscopy (SEM) and Fourier Transform Infrared Spectrophotometer (FTIR), respectively. Moreover, the water retention studies was conducted in order to investigate the efficiency of CRWR coated with SAP and carbonaceous-SAP in retaining the water content in different soil (organic and top soil). Based on the results, the CRWR fertilizer that was coated with carbonaceous-SAP had higher water absorbency value than the CRWR fertilizer without carbonaceous-SAP. Meanwhile, 1 wt % CRWR fertilizer in organic soil gave higher water retention ability compared with 1 wt % CRWR fertilizer in top soil. In conclusion, CRWR fertilizer coated with carbonaceous filler gave a significant influence on the water absorbency and in controlling the nutrients release rate as well as function as water retention in soil.

1 Introduction

The agriculture sector plays an important role in economic development, uplifting rural incomes and help in ensuring national food security [1]. Internationally agriculture has become the centre of cutting-edge reaserach and development (R&D) as the drive to feed and cope with global population. In general, the main factors that affect the growth of plants and their quality are quantity of water and fertilizers that can be absorb by plants. Besides maintaining the water resources, the used of fertilizer is also crucial as it function as one of the nutrients sources for plant in order to achieve maximum efficiency and highest quality product. In order to increase the production, excessive fertilizers are normally were practiced in agricultural activities in order to fulfill and the required nutrients as it will subsequently affect the quality of yield [2]. This practice however is not efficient as not all the nutrients from these fertilizers will be absorbed by the plant because it will be washed or leached out during rain [3]. It has been reported by Al-Zahrani [4] that only 30% of the fertilizers are used and utilize by plants while the balance is lost to environment. In order to increase the production, excessive fertilizers are normally were practiced in agricultural activities in order to fulfill and the required nutrients as it will subsequently affect the quality of yield [2]. Beside of the increasing of productive cost, excessive use of fertilizers also causes a negative impact especially on the environment because some fertilizers does contains heavy metals such as cadmium, chromium and high concentrations of radionuclides [2].

The loss of fertilizer will cause large economic losses and environmental pollution [5]. One of the solutions is by applying controlled release fertilizer in agricultural activities to enhance the quality and yield of product. Recently, the use of slow release fertilizer is a new trend to save fertilizer consumption and to minimize environmental pollution [6]. Practically, the factors that controlling the release of the fertilizers include temperature, microbial activity, soil moisture, pH and organic matter. The present study proposed a fabrication of controlled release water retention (CRWR) fertilizer that coated with superabsorbent polymer material. CRWR fertilizer is known as fertilizer that was coated with superabsorbent polymer (SAPs) which was design to control the nutrients release from fertilizer and enhance the efficiency of nutrients use for plant growth [7]. According to Liu & Guo [8], SAPs can be defined as a long chains polymer which are slightly cross-linked that has ability to swell and maintain bulky of water and aqueous solutions after absorb. In agricultural uses, higher water-holding capacity and nutrient retention in soils is important as it can increase the soil's aeration and microbial activity, reduce the environmental impact from water-soluble fertilizers and lower frequency of irrigation [9].

Moreover, the use of biochar for agricultural application had attracted many attention and recent studies have focused on the application of biochar to soils for carbon storage and soil fertilization [10]. The application of biochar has a number of specific function in the natural environment as it is beneficial to prevent global warming and increases the functionality of soils. Carbon from biochar have the ability to improve the physical properties; by increasing the cation exchange capacity (CEC) and water-holding capacity, of soil [11]. Pyrolysis is the best known method to produce biochar as it is much environmental friendly. Pyrolysis of the biomass feedstocks enables the biomass conversion to biochar whose subsequent application to soil is in a more stabilized form [12]. The pyrolysis process requires less of the supplied fuel than traditional combustion.

This present study reported the pyrolysis process of biochar, solution polymerization process of SAPs and carbonaceous-SAPs to be coated as CRWR fertilizer, and influence of these coating materials on the water absorbency, release behavior and water retention properties of the CRWR fertilizer in different types of soils.

2 Materials and methods

2.1 Pyrolysis of carbonaceous particles

Pyrolysis of carbonaceous particles was carried out using biomass of empty fruit bunch (EFB) using Nabertherm furnace with nitrogen gas. During the process, the nitrogen gas flow rate was set at 250 ml/min to purge out the presence of oxygen inside the furnace. This pyrolysis process was carried out at 350°C for 1 h with heating rate of 10°C/min. The obtained carbonaceous particles were collected and characterized for moisture content, ash content, volatile content, and calorific value.

2.2 Synthesis of SAP and carbonaceous-SAP

In this study, acrylic acid (AA) and acrylamide (AM) that acts as monomer and co-monomer, respectively were neutralized with NaOH and distilled water in 250 mL three neck flasks equipped with stirrer, condenser, nitrogen line and thermometer. MBA acts as a cross-linker agent was added in the monomer solution and was stirred at room temperature until the MBA was completely dissolved. The flask then was immersed in a water bath and heated up to 70 °C and then APS initiator was added to the solution that was magnetically stirred at 300 rpm [13]. The reaction mixture was left to stir for 2 hours until gel point before the resulting hydrogel was removed from the reaction and dried overnight in oven at 60 °C. In prepration of carbonaceous-SAP, same procedure was repeated with addition of 0.5 wt% of carbonaceous particles that obtained from pyrolysis process was used as filler. Finally, the commercial fertilizer granule were dipped in starch solution and then immediately placed on the prepared SAPs and carbonaceous-SAPs powder to form CRWR fertilizer. The CRWR fertilizer then was dried in oven at 60 °C to obtain the final product of CRWR fertilizer.

2.3 Water Absorbency Properties

In order to determine the water absorbency properties, about 2 g of sample of SAPs and carbonaceous-SAPs were put in a tea bag (M_0) was immersed in 500 mL tap water and allowed to soak at room temperature up to 7 h. Then, the immersed tea bags were taken out from water for every certain time and weighted (M_1). The water absorbency (WA) was calculated using the Equation 1 [14]:

$$WA = \frac{M_1 - M_0}{M_0} \qquad (1)$$

2.4 Water Retention Properties of CRWR in soil

The water retention of CRWR fertilizer in soil was determined by mixed the CRWR fertilizer in soil for a certain period of time. In this step, 2 g of CRWR fertilizer was well mixed with 200 g of soil (ratio 1:100) and kept in a beaker. Then 120 mL of tap water was slowly added into the beaker and weighed (W_1) and maintained at room temperature. In this step, the control (fertilizer without any coating material) was also being observed for comparison purposed. The beakers that contained fertilizer and soil were weighed every 2 days (W) over a period of 21 days. The water retention ratio ($WR\%$) of soil was calculated using the Equation 2 [15]:

$$WR\% = \frac{W}{W_1} X100 \qquad (2)$$

2.5 Fourier Transform Infrared (FTIR) Spectrophometer

The chemical structure of the SAP samples was established using FTIR spectroscopy. FTIR spectra (thermo, Nicolet, NEXUS, TM) of the samples were taken in KBr pellets.

DOI: 10.1051/matecconf/20166906001

2.6 Morphological Properties of SAP and CRWR

The surface morphology of the sample was observed by using Field Emission Scanning Microscope (JSM-5610 LV SEM). The sample is sputter-coated with a gold layer in a vacuum using an Emitech K550 Sputter Coater, prior to analysis.

3 Result and discussions

3.1 Pyrolysis of carbonaceous particles

The efficiency of any future energy application process depends on the calorific value of the inlet, where high calorific value will increase the amount of energy released. In addition to calorific value, other important properties are moisture content, ash content and volatile content. Energy source that have high calorific value, low moisture content and low ash content can be burned without additional fuel, whereas materials with low calorific value, high moisture content and high ash required supplementary fuel. Analysis to determine the calorific value, moisture content, ash content and volatile content is called a proximate analysis. Fig. 1 depicted proximate analysis of carbonaceous particles that obtained from pyrolysis process. The decomposition mechanism of pyrolysis involves a significant dehydration as the changes in the H:C and O:C atomic ratios of biomass follows the dehydration pathway. The higher loss in oxygen and hydrogen compared to carbon is highly related to the increase in energy value of the biomass. Compared to oxygen and hydrogen, the bonds between carbon atoms are more stable. Therefore, more energy is needed to break the carbon bond, which will reduce the ability of the combustion.

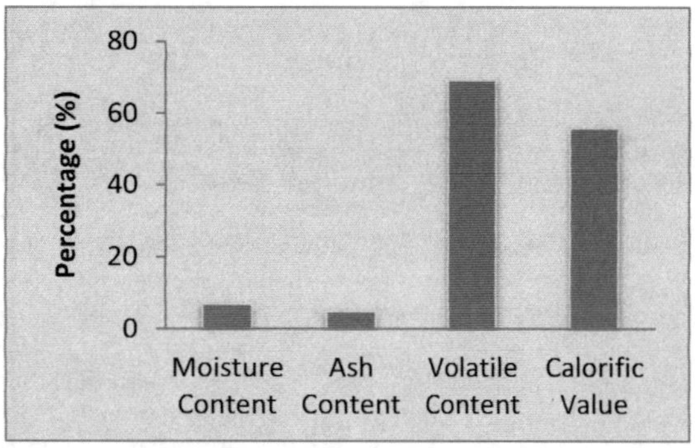

Figure 1. Moisture content, ash content, volatile content and calorific value of carbonaceous particles.

3.2 Water Absorbency Properties

The water absorbency of SAPs and carbonaceous-SAPs at various swelling time until 420 min is shown in Fig. 2(a). The absorption value of both samples (SAPs and carbonaceous-SAPs) were increased tremendously at the beginning of swelling time and reached a maximum value at certain period. From Fig. 2(a), it can be seen that the carbonaceous-SAPs had higher water absorbency value than the SAPs throughout the swelling time. The SAPs sample reached saturation or maximum absorbency value of 25 g/g at 180 min (3 h) whereas the carbonaceous-SAPs reached maximum absorbency value of 40 g/g at 300 min (5 h). From this observation, it can be stated that the addition of carbonaceous particles improved the water absorbency of the samples but somehow reduce the absorbency rate as it required longer swelling time to achieve maximum absorption compare with the SAPs sample.

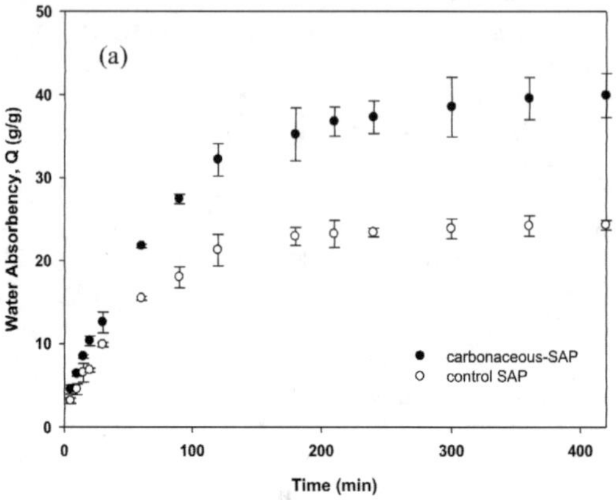

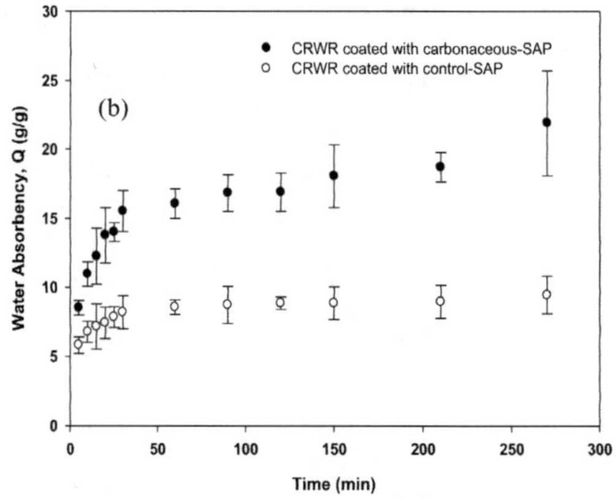

Figure 2. Water absorbency of (a) SAP and carbonaceous-SAP, (b) CRWR coated with SAP and carbonaceous-SAP.

According to Zangi *et al.*, [16], urea fertilizer is easily dissolve in water and normally takes around 270 seconds to completely dissolve in water. Fig. 2(b) showed water absorbency of CRWR fertilizer coated with SAPs and carbonaceous-SAPs with time. From Fig. 2(b), it is clearly observed that CRWR fertilizer that coated with carbonaceous-SAPs show the higher water absorption value with slow water absorbency rate than CRWR coated with SAPs. CRWR fertilizer coated with SAPs achieved maximum point faster than the fertilizer with carbonaceous filler which is about 8 g/g at 1 h meanwhile the CRWR fertilizer coated with carbonaceous-SAPs approached maximum value about 18 g/g at 150 min or 2 h 30 min. This observation can be related to the

DOI: 10.1051/matecconf/20166906001

composition of coating layer. It was believed that the addition of carbonaceous filler that has hydrophilic properties increase the water uptake of the coating material. At the same time, the presence of this carbonaceous particles also act as barrier that slow down the diffusion rate of the water. The observation of water absorbency of CRWR fertilizers can be related to its morphology structure. Porous structure of carbaceous particles also results in the increase of water absorbency value. The presence of pores within the CRWR fertilizer coated with carbonaceous-SAPs allow the water to diffuse in and occupy the space [17]. This eventually cause the CRWR fertilizer coated with carbonaceous-SAPs can store large amount of water compared to the CRWR fertilizer coated with SAPs.

3.3 Water Retention Properties of CRWR in soil

Some of the important soil physical conditions that influence the crop production are water capacity and water retention properties. The incorporation of carbonaceous particles could improved soil physical properties by increase aggregate stability and reduce bulk density which eventually affects the water retention properties of the soil.

Table 1. Comparison of Water Retention Properties of 1 wt.% CRWR Fertilizer Coated Carbonaceous-SAP in Organic and Top Soil.

Day	Water Retention (%)	
	CRWR in organic soil	CRWR in top soil
1	100.00	100.00
3	94.24	92.28
5	89.24	85.66
7	84.03	79.04
9	79.64	73.11
11	75.71	67.88
13	71.68	62.65
15	67.61	57.12
17	64.15	55.56
19	62.11	51.94
21	60.04	48.33

Organic soil normally used as potting soil for plantation. It is the organic matter component of soil, consisting of plant and animal residues at various stages of decomposition, cells and tissues of soil organisms, and substances synthesized by soil organisms [18]. Meanwhile, the top soil in general is the top layer of the earth's surface. It is in dark brown colour and high in organic matter, which makes it very easy to till and fertilize ground for growing plants. It is comprised of the three main groups of organic matter which are humus, clay particles and sand. In this study, the effect of 1 wt. % of CRWR fertilizer coated with carbonaceous-SAPs in two different soils on water retention properties were observed and summarized in Table 1. The obtained result in Table 1 shows that both CRWR coated with SAPs and carbonaceous-SAPs have capability to absorb and stored water before slowly released the nutrients with soil moisture [6]. By comparing the results in Table 1, it can also be seen that the water retention of CRWR fertilizer coated with carbonaceous-SAPs in top soil decreased faster than the organic soil. In 7th day, the water retention of CRWR fertilizer in organic soil and top soil is 84.03 % and 79.04 %, respectively. This results shows that the organic soil had higher water retention ability for CRWR fertilizer compared with top soil. According to El-Rahim [19], the water retention ability in soils is related to the way of soil in holding the water. There is two ways for soil holding water which are a film coating on soil particles and in the pore space between particles. Moreover, the soil porosity also could affect the water retention in soil. Large pore spaces between the soils could prevent water retention, where the water transpiration is quickly. This soil porosity are depends on soil texture and structure, where water can be held tighter in small pores than in large pores through capillary forces [19]. Therefore, organic soils with small pores can hold more water than top soil.

3.4 Fourier Transform Infrared (FTIR) Spectrophometer

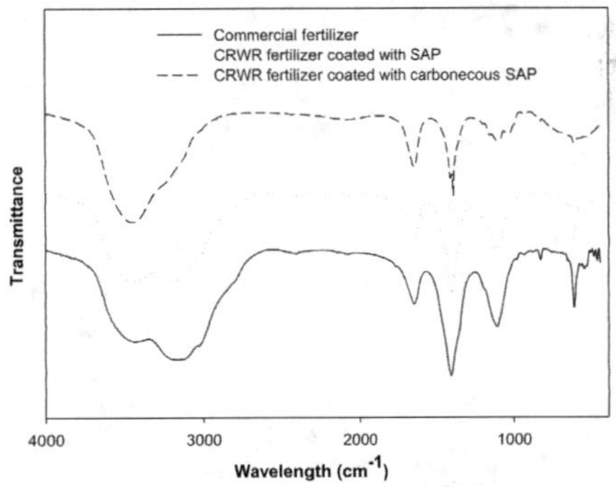

Figure 3. The IR spectrum of commercial fertilizer, CRWR fertilizer coated with SAP and CRWR fertilizer coated with carbonaceous-SAP.

Fig. 3 shows the peak of O-H, C=C, C-H and C-O bonds for commercial fertilizer and CRWR fertilizers coated with SAPs and carbonaceous-SAPs. Based on the Fig. 3, the commercial fertilizer had the highest wavelength peak at 3163 cm^{-1} which represents the carboxylic group, and the peaks at 1636, 1402 and 1107 cm^{-1} indicate the presence of alkene, alkane and alcohol, phenol or ester groups respectively. Meanwhile, the spectrum for the CRWR fertilizer coated with SAP and carbonaceous-SAPs shows the highest wavelength at 3436-3437 cm^{-1} representing O-H bond, at 1638 cm^{-1} representing alkene group, and at 1384-1401 cm^{-1} representing alkane. Moreover, for the CRWR fertilizer coated with carbonaceous-SAPs, there is three wavelengths are close to one another at 1151, 1104 and 1023 cm^{-1} that corresponding to the C-O group from the carbonaceous filler. In this study, there is an intermolecular hydrogen bond formation between the fertilizer core and SAP monomers. The broad stretching band of hydroxyl groups

(3600-3300 cm^{-1}) is relating to the stretching O–H from the intermolecular and intramolecular hydrogen bonds.

3.5 Morphological Properties of SAP and CRWR

The surface and cross sectional image of the two samples which were CRWR fertilizer coated with SAPs and carbonaceous-SAPs that were observed under SEM at magnification of 50x are shown in Fig. 4. Based on Fig. 4, the morphology of surface of both CRWR fertilizer coated with SAPs and carbonaceous-SAPs are loose and porous, with presence of holes and cracks of varying sizes which allowing the diffusion of water into the structure. According to Wu and Liu [15], CRWR fertilizer with course and porous surface could help the sample to absorb water quickly in large amount. The cross-section for the both CRWR fertilizers under SEM also reveal the presence of two layers as shown in Fig. 4. However, in this study, the presence of this two layers are not uniform in thickness due to the variety of SAPs powder used during the coating process.

Figure 4. SEM of the cross-section of CRWR fertilizer coated with (a) SAP (b) carbonaceous-SAP.

4 Conclusion

In this study, the single coated of controlled release water retention (CRWR) fertilizer has been successfully synthesized via solution polymerization method. The presence of hydrophilic carbonaceous particles increased the water absorbency of the sample thus results in higher water absorbency of CRWR fertilizer coated with carbonaceous-SAPs. The obtained CRWR fertilizer coated with SAPs and carbonaceous-SAPs are loose and porous, with presence of holes that allowing the entry of water into the system and expands their size due to the storage of water at its outer layer. Result from FTIR analysis shows the presence of O-H, C=C, N-H, C-H, C-O, C-S and O=C=O bonds that results in high water absorbency of CRWR fertilizer. Moreover, 1 wt % CRWR fertilizer in organic soil had higher water retention ability compared with 1 wt % CRWR fertilizer in top soil. In conclusion, CRWR fertilizer gave a significant influence on the performance of water absorbency and controlled of nutrients release rate as well as function as water retention in soil.

References

1. Istikoma, Quratul-Ain, A. D. Abdul Rahman, (2015). The Transformation of Agriculture Based Economy to an Industrial Sector through Crowd Sourcing In Malaysia, *International Journal of Computer Science and Information Technology Research,* vol. 3, Issue 1, pp34-41.
2. S. Savci, (2012). An Agricultural Pollutant: Chemical Fertilizer. *International Journal of Environmental Science and Development.*
3. Y. Munusamy, Y. T. Jeng, C. S. Yong, (2013). Development of controlled release fertilizer using biodegradable superabsorbent polymers for plantations in Malaysia.
4. S. Al-Zahrani, (1999). Controlled-release of fertilizers: modelling and simulation. *International Journal of Engineering Science*, vol. 37, no. 10, pp.1299-1307.
5. M. Guo, M. Liu, Z. Hu, F. Zhan, L. Wu, (2005). Preparation and properties of a slowrelease NP compound fertilizer with superabsorbent and moisture preservation, *Journal of Applied Polymer Science*, pp2132–2138.
6. L. Wu, M. Liu, R. Liang, (2008). Preparation and properties of a double-coated slow-release NPK compound fertilizer with superabsorbent and water-retention. *Bioresource Technology* 99, pp547-554.
7. A. Shaviv, (2005). Controlled release fertilizers. *IFA International Workshop on Enhanced-Efficiency Fertilizers.*
8. M. Liu, T. Guo, (2001). Preparation and Swelling Properties of Crosslinked Sodium Polyacrylate, *Journal of Applied Polymer Science* **82**, pp1515-1520.
9. H. A. El-Rahim, E. S. Hegazy, H. El-Mohdy, (2004). Radiation synthesis of hydrogels hydrogels to enhance sandy soils water retention and increase plant performance, *Journal of Applied Polymer Science*, pp.1360–1371.
10. B. Glaser, (2002). Ameliorating Physical and Chemical Properties of Highly Weathered Soils in the Tropics with Charcoal: A review, *Biological and Fertiliser of Soils, ISSN0178-2762, 35(4)* , 219-230.
11. S. Jamari, S. Ghazali, W. Yaacob, (2015). Effect of Superabsorbent Polymer Composite Filled Carbon

Fiber Towards the Germination of Abelmoschus Esculentus, *JOAAT*, **2** (2), pp156-159.

12. H. K. Nsamba, (2015). Sustainable Technologies for Small-Scale Biochar Production : A Review, *Journal of Sustainable Bioenergy Systems, 5.* , 10-31 .
13. N. Mohamad, A. H. Nor Nadiah, A. R. Jeefferie, D. Mohd Fairuz, (2011). Effect of Chitosan Gelatinization Temperature on Water Absorption and Water Retention of Chitosan Based Urea Fertilizer.
14. A. B. Pitaloka, A. H. Saputra, A. H., M. Nasikin, (2013). Water Hyacinth for Superabsorbent Polymer Material. *World Applied Sciences Journal*, pp747-754.
15. L. Wu, M. Liu, (2008). Preparation and properties of chitosan-coated NPK compound fertilizer with controlled-release and water-retention. *Carbohydrate Polymers*, **72** (2), pp240-247.
16. R. Zangi, R. Zhou, B. J. Berne, (2009). Urea's action on hydrophobic interactions. *Journal of American Chemical Society*, **131**: pp1535-1541.
17. Z. Guohua, L. Ya, F. Cuilan, Z. Min, Z. Caiqiong, C. Zongdao, (2006). Water resistance, mechanical properties and biodegradability of methylated-cornstarch/poly(vinyl alcohol) blend film, *Polymer Degradation and Stability*, **91**,pp703-711.
18. N. C. Brady, R. R. Weil, *The nature and properties of soil*, 12th ed. Prentice Hall, Inc., Upper Saddle River, NJ., 1999.
19. H. A. El-Rahim (2006). Characterization and possible agricultural application of polyacrylamide/sodium alginate crosslinked hydrogels prepared by ionizing radiation, *J Appl Polym Sci* **101**: pp3572-3580.

Thermal behavior and the solvent effects of ρ-Methoxy Tetraphenylporphyrin (TOMPP), Copper Porphyrin (CuTOMPP), and Nitroporphyrin (CuTOMPP-NO_2)

Tossapon Phromsatit, Wootthiphan Jantayot, Kusuma Pinsuwan, Autthavit Nuchthanom and Supakorn Boonyuen

Chemistry Department, Thammasat University, Pathumtani, 12120 Thailand

Abstract. In this work, ρ-methoxy tetraphenylporphyrin (TOMPP) Copper porphyrin (CuTOMPP) and nitro porphyrin (CuTOMPP-NO_2) are synthesized following Adler Longo method by the reaction of pyrrole with and ρ-anisaldehyde. These compounds have been characterized by UV-Vis, fluorescence spectroscopy and thermogravimetric analysis (TGA). The ρ-methoxy tetraphenylporphyrin (TOMPP) was obtained in 26% yield. Reactions of TOMPP with copper ions gave CuTOMPP in 85% yield. Moreover, the nitration of CuTOMPP gave CuTOMPP-NO_2 in 0.03% yield. The absorption spectra for TOMPP exhibited one S-band (416 nm) and four Q-bands (518, 555, 593 and 650 nm) in dichloromethane solvent. The absorption band of copper porphyrin and nitroporphyrin displays one S band (415 and 423) and two Q bands (range from 550 – 600 nm). When excited to 530-570 nm, the fluorescence spectra show only one emission band at 645-654 nm for TOMPP and copper complexes. The results confirmed the solvent effect on the electronic absorption as found a small red shift from 414 nm (chloroform) to 417 nm (ethyl acetate) for TOMPP. The thermal behaviors including the possible phase transition of porphyrins were studied during the heating process at temperature of 298-973 K. Copper complexes exhibits the thermal stability from 653.3 K, which is slightly lower than that of the free base porphyrin ligand (TOMPP 703.9 K).

1 Introduction

Porphyrins were heterocyclic macrocycle organic compounds which were found in nature. Naturally occurring porphyrins played an important in the maintenance of the animal and plant bodies [1, 2]. Hemoglobin, cytochromes and chlorophylls were the best known porphyrin pigments [3, 4]. Hemoglobin and cytochromes contain iron(II) ion and chlorophylls contained magnesium(II) ion in the metal center hole [5, 6].

Free base porphyrins are known as substances with a great capability to form highly stable metalloporphyrins, having a great potential as corrosion inhibitors. Their planar molecules containing four pyrrole subunits present bonding sites for complexes formation at the nitrogen atoms. The whole molecule structure shows conjugated double bonds due to the -CH= bridging groups and thus, a very mobile electronic system. These properties are very important in the adsorption process of these compounds on diverse metals surfaces [7].

Synthesis of porphyrins and their metal complexes have received much attention. This has been mainly due to the use of these compounds in catalysis [8], as materials with novel electrical properties [9] and as biomimetic model systems of primary processes of natural photosynthesis [10]. They can be used as photosensitizing drugs in photodynamic therapy (PDT). Photodynamic therapy is emerging as an important treatment for many diseases. Many applications of PDT involve killing undesirable disease-causing cells such as malignant cancer cells or pathogenic microorganisms [11]. PDT is also used for destroy unwanted tissues such as tumors [12], new blood vessels [13], and atherosclerotic plaques [14]. An extra stable porphyrin occurs by complexation with transition metal ions explained by the macrocyclic effect [15]. Porphyrins and their derivatives have well-known technological applications. Concerning to it can be applied as dyes in solar cells to improve efficiency [16]. Porphyrins have the characteristic properties and can be applied in many applications such as molecular electronic devices.

In the present work, the application of Adler Longo synthesis method was selected to study for the free base porphyrins ligand (TOMPP). The ρ-Methoxy tetraphenylporphyrin (TOMPP) and metalloporphyrins were prepared and purified polar characterization. The complexes were characterized by IR spectroscopy, ^{1}H and ^{13}C NMR spectroscopy, mass spectrometry (MS), CHN analysis, thermogravimetric analysis (TGA), UV-Vis and fluorescence spectroscopy. To study the effect of solvent, the absorption spectra and emission spectra of free base porphyrin were analyzed by using UV-Vis spectrophotometry and spectrofluorometer, in dichloromethane, toluene, ethyl acetate and chloroform. The thermal gravimetric analysis results would represent

DOI: 10.1051/matecconf/20166906002

the stability of complexes, when it was prepared in real application such as coating on ITO glass.

2 Experimental

2.1 Measurements

Properties of ρ-methoxy tetraphenylporphyrin (TOMPP), copper porphyrin (CuTOMPP) and nitro porphyrin (CuTOMPP-NO_2) were characterized by UV-Vis spectroscopy in the wavelength range of 200-800 nm and fluorescence spectroscopy in the wavelength range of 530-700 nm. UV-Visible spectra were obtained on Shimadzu UV-spectrometer (UV-1700). Fluorescence spectra were recorded on Jasco spectrofluorometer (FP-6200). The FT-IR spectra were recorded on Perkin Elmer infrared spectrophotometer (spectrum GX). ^{1}H and ^{13}C NMR spectra were recorded with a Bruker (FT-NMR advance 400 MHz) spectrometer in chloroform-d with tetramethylsilane (Me_4Si) as an internal standard. Mass spectra were obtained on Thermo Finnigan (LCQ Advantage) using CH_2Cl_2. CHN analysis was carried out on Perkin Elmer (2400) elemental analyzer. TGA were recorded on Perkin Elmer (TGA7).

2.2 Synthesis of p-methoxy tetraphenyl porphyrin, its metal complexes, and nitro derivative

General procedure: The ρ-methoxy tetraphenylporphyrin (TOMPP), were synthesized by refluxing pyrrole with ρ-anisaldehyde in propionic acid. The product has methoxy groups on para position of each TPP ring, as shown in Fig. 1. The metalloporphyrin were prepared by refluxing TOMPP ligand with metal acetate in DMF (Fig. 2). Lastly nitro group have been insert in CuTOMPP via reaction between CuTOMPP and copper nitrate (Fig. 3.). The crude products were purified by silica column chromatography with dichloromethane and hexane. There derivative were synthesized similarly.

Synthesis of ρ-methoxy tetraphenylporphyrin (TOMPP): The procedure was started by refluxing pyrrole (1 mL, 14.3 mmol) and ρ-anisaldehyde (1.5 mL, 15.6 mmol) in propionic acid (100 mL). The reaction was stirred at 363 K for 1 h. The title compound was isolated by column chromatography (silica gel, hexane: dichloromethane) as a purple solid (26% yield). IR (NaCl): 3435, 3053, 2987, 1606, 1550, 1265, 1163, 1017, 895 cm^{-1}. ^{1}H-NMR (400 MHz, $CDCl_3$): δ8.88 (8H, Pyrrole, β-H), 8.13 (8H, Phenyl,o-H), 7.35 (8H, Phenyl, m-H), 4.22 (12H, -OCH_3), -2.82 (2H, Pyrrole, N-H). ^{13}C-NMR (100 MHz, $CDCl_3$): δ128.79 (Cmeso), 159.45 (Cα, pyrrole ring), 130.81(Cβ, pyrrole ring), 135.56, 134.87 (-C_6H_4), 134.72, 130.81(phenyl ring), 55.56 (-OCH_3). Elemental analysis calcd (%) for $C_{48}H_{38}N_4O_4$ (734.84): C 78.45, H 5.21, N 7.62; found: C 76.42, H 5.75, N 7.42. The lower carbon contents many related to solvent in the compound, which need further characterization. Mass m/z (ESI): found 735.38.

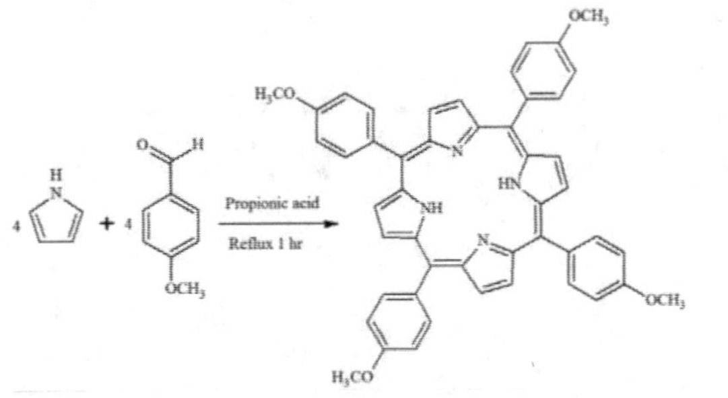

Figure 1. Synthesis of ρ-methoxy tetraphenylporphyrin (TOMPP).

The ρ-methoxy tetraphenylporphyrin (0.1 g, 0.14 mmol) and copper acetate (0.1 g, 0.55 mmol) were heated under refluxed in DMF (20 mL) The CuTOMPP was purified by column chromatography (silica gel, hexane: dichloromethane) and gelded as a purple solid (85% yield). ^{1}H-NMR (400 MHz, $CDCl_3$): δ8.88 (8H, Pyrrole, β-H), 8.14 (8H, Phenyl,o-H), 7.35 (8H, Phenyl, m-H), 4.21 (12H, -OCH_3). IR (NaCl): 3055, 2987, 1604, 1551, 1262, 1150, 896 cm^{-1}. Elemental analysis calcd (%) for $CuC_{48}H_{36}N_4O_4$ (796.37): C 72.39, H 4.56, N 7.04; found: C 71.78, H 4.65, N 7.02. Mass m/z (ESI): found 797.46.

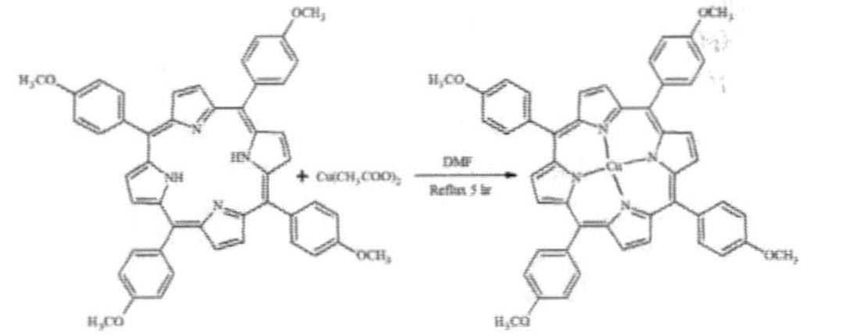

Figure 2. Synthesis of copper porphyrin complex (CuTOMPP)

Synthesis of 2-nitro-methoxy tetraphenylporphyrinato copper(II) (CuTOMPP-NO_2): The general procedure was magnetically stirred CuTOMPP (0.2 g, 3 mmol) and excess copper nitrate in chloroform (100 mL) and acetic anhydride (9 mL) for 48 hours at room temperature. The title compound was isolated by column chromatography (silica gel, hexane: dichloromethane) as a purple solid (0.03% yield) ^{1}H-NMR (400 MHz, $CDCl_3$): δ7.26 (20H, Phenyl), 3.98 (12H, -OCH_3). IR (NaCl): 2928, 2831, 1606, 1501, 1463, 1340, 1248, 1174, 800 cm^{-1}. Mass m/z (ESI): found 842.65.

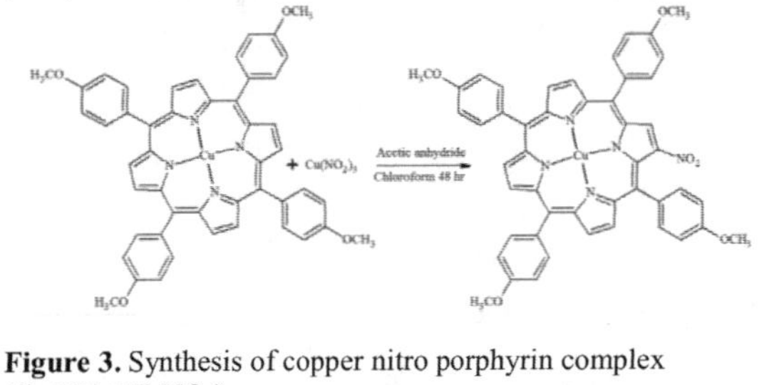

Figure 3. Synthesis of copper nitro porphyrin complex (CuTOMPP-NO_2)

3 Results and discussion

DOI: 10.1051/matecconf/20166906002

Table 1. The absorption data of TOMPP in various solvents.

Solvent	TOMPP				
	Soret band (nm), ε ($M^{-1}cm^{-1}$)	Q band (nm), ε ($M^{-1}cm^{-1}$)			
Dichloromethane	416, 29129	518, 5414	555, 3793	593, 1778	650, 2400
Toluene	416, 21088	518, 2881	555, 1962	596, 856	653, 1255
Ethyl acetate	417, 33034	516, 6468	558, 4565	594, 1976	651, 2750
Chloroform	414, 39198	519, 8313	556, 5683	594, 2770	651, 3429

* All solution was prepared in the concentration of $1x10^{-4}$ mol/L, (n =3, %RSD ≤ 1.4)

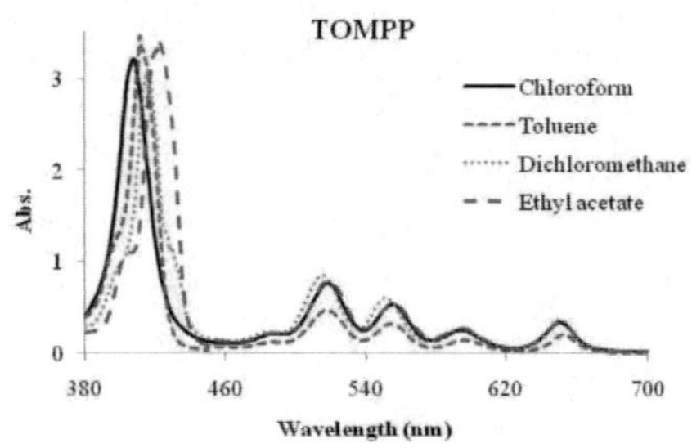

Figure 4. UV-Vis spectra of TOMPP in various solvents

Table 2. The absorption data of TOMPP, CuTOMPP, and $CuTOMPP-NO_2$ in dichloromethane.

Porphyrins	Dichloromethane				
	Soret band (nm), ε ($M^{-1}cm^{-1}$)	Q band (nm), ε ($M^{-1}cm^{-1}$)			
TOMPP	416, 404283	518, 15986	555, 10937	593, 5468	650, 6731
CuTOMPP	415, 510297	-	541, 7993	585, 5468	-
$CuTOMPP-NO_2$	423, 224227	-	551, 17248	597, 13882	-

*All solution were prepared in the concentration of 2.38 x 10^{-6} mol/L, (n =3, %RSD ≤ 1.6)

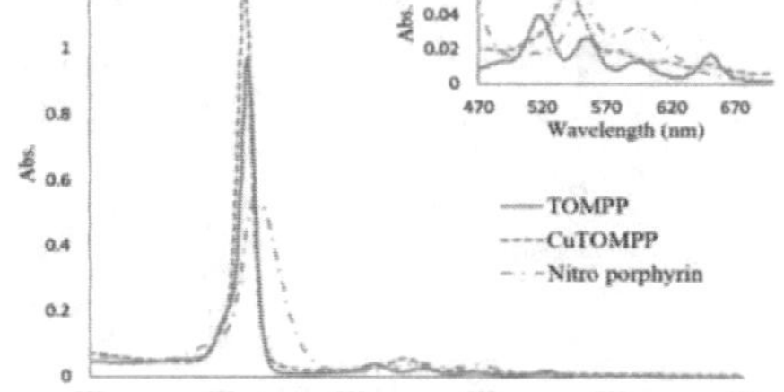

Figure 5. UV-Vis spectra of TOMPP, CuTOMPP, and $CuTOMPP-NO_2$ in dichloromethane

The intense colour of porphyrins was derived from the highly conjugated π-electron systems. The most attraction feature of porphyrins was their characteristic UV-Visible spectra that consist of two distinct regions; in the near ultraviolet is called the Soret band (380-450 nm) and in the visible region is called Q-band (500-700 nm). The absorption spectra of the free base porphyrin showed significant characteristics of one soret band and four Q-bands, as shown in Fig. 4. The electronic spectra of the TOMPP in various solvents were investigated and the λ_{max} values were given in Table 1. In metalloporphyrin, the proton on NH group of porphyrin were deprotonated and then nitrogen atom binds with metal ion to give the metal-porphyrin. The metal ions acting as Lewis acids interact with the lone-pair electrons of porphyrin ligand. The intense absorption band, related due to absorption(s) with in the porphyrin ligands, involving the excitation of electrons from π to π* prophyrin ring orbital [17], as shown in Fig. 5. The absorption band of metalloporphyrin displays only two Q-bands. The results were shown in Table 2 due to difference energy levels.

The solvent are effected on electronic absorption of TOMPP, as found a small red shift from 414 nm (chloroform) to 417 nm (ethyl acetate). The lower energy gap found in ethyl acetate solvent are related to the smaller HOMO-LUMO gap of the complexes. The further information for solvent effect on metalloporphyrins are needed to be studied. Similar result on the solvent on lower Q-band intensity for metalloporphyrins are found and reported as solvatochromism. Those complexes show linear correlation between polarity of solvent and dipole moments of complexes [18]. However, those results confirmed the solvent effect on the electronic absorption. Furthermore the weak coordination between each metalloporphyrin and solvent were clearly found. The sign of solvatochromism depends on the variation of the dipole moment of complexes and the absorption spectra with solvent polarity. When comparing with free base porphyrin (TOMPP) with metal complexes the soret band was little shifted while the Q-bands showed only 2 band Q-bands (Table 2). The absorption spectrum of Q-bands was changed relating to symmetry effect. The change in the spectrum (fewer peak in Table 2) on metal added is due to increased symmetry relative to the free base porphyrin similar to the previous reported by Boudif [19]. The two hydrogen atoms on the nitrogen atom in the free base porphyrins reduce the ring symmetry.

Fluorescence spectroscopy of TOMPP and metalloporphyrins were characterized in dichloromethane and the results were shown in Table 3. The enlargement of the π-conjugation yields the emission characteristics at the emission range from 645-654 nm when excited at 530-570 nm. This is manifested in a blue shift of the emission maximum compared to the free base ligand as the π-delocalization increases. This finding is consistent with the similar trend observed in the absorption spectra as above.

The term of band gap refers to the energy difference between the top of the valence band to the bottom of the conduction band, electrons are able to jump from one band to another. In order for an electron to jump from a valence band to a conduction band, it requires a specific minimum amount of energy for the transition, called the band gap energy [20]. The estimated energy gap determined from an intersection of UV-Vis absorption (Q4-band) and fluorescence emission spectrum [21]. The calculated energy gap of TOMPP was 1.902 eV. The intersection of spectra of TOMPP showed in Fig. 6.

Table 3. The excitation and emission wavelength of TOMPP and copper complexes.

Porphyrin	$CH2Cl_2$		Toluene		Ethyl acetate		$CHCl_3$	
	Ex.	Em.	Ex.	Em.	Ex.	Em.	Ex.	Em.
TOMPP	530	654	530	655	530	654	530	652
CuTOMPP	570	654	570	656	560	571, 654	560	653

All solution were prepared in the concentration of 1 x 10^{-4} mol/L, (n=3, %RSD ≤ 1.5)
*Ex.; Excitation wavelength (nm), Em.; Emission wavelength (nm)

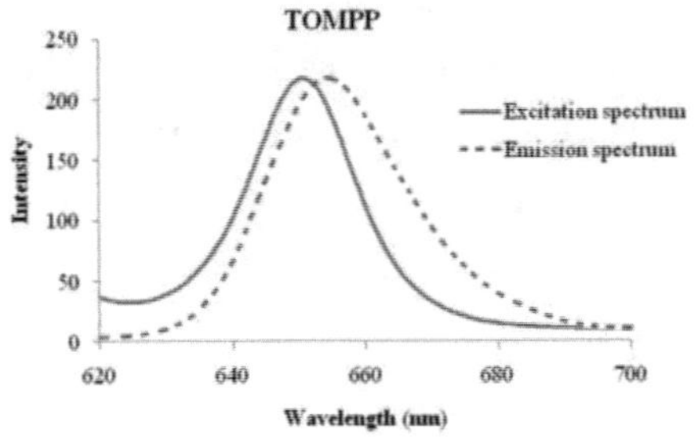

Figure 6. Excitation spectrum and emission spectrum of TOMPP in dichloromethane

Under the conditions that the thermogravimetry measurements were performed (open cells and nitrogen flow), the first observation is that decomposition temperature of the free base porphyrin TOMPP, is really high compared to metalloporphyrin. The condensation of pyrrole with *ρ*-anisaldehyde will afford the remarkable stability to TOMPP, which decompose at T = 703.9 K. The greater thermal stability may be related to the bond between metal ion and ligand in complexes and be a good choice. The metalloporphyrin complexes show the greater thermal stability from Cu-TOMPP (shown in Table 4 and Fig. 7.). Consider with TGA, metal complexes were interesting to use in alternative energy of solar cell [22]. However, the metal complexes show the lower thermal stability than pure ligand that may be related to weaker bond in metal complexes.

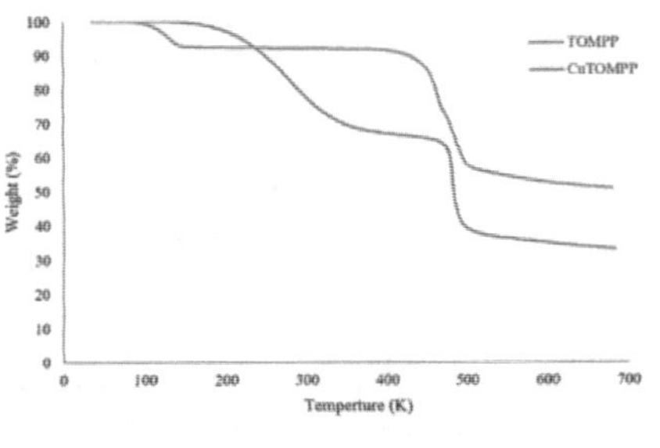

Figure 7. Themogravimetry curve of TOMPP and CuTOMPP

Table 4. Temperture of decomposition of TOMPP and CuTOMPP.

Compounds	I		II	
	T decomp (K)	Weight loss%	T decomp (K)	Weight loss%
TOMPP	443.2	31.3	703.9	29.8
Cu-TOMPP	348.4*	7.9	653.3	37.3

*water vapor

4 Conclusions

ρ-Methoxy tetraphenylporphyrin (TOMPP) and metalloporphyrins have been successfully synthesized and their identities were confirmed by mass spectrometry (MS), CHN elemental analysis, ^{1}H, ^{13}C NMR, FT-IR spectra, thermogravimetric analysis (TGA), UV-Vis and fluorescence spectroscopy. The free base ligand (TOMPP) at optimal condition (refluxing in propionic acid) was obtained 26% yield. Reactions of TOMPP with metal ions (Cu^{2+}) gave CuTOMPP in 85% yield. The nitration reaction provided CuTOMPP-NO_2 in 0.03% yield. The UV-Vis absorption spectra for TOMPP exhibited a single S-band with four Q-band, which the metalloporphyrin complexes and insertion nitrocomplex show a single S-band couple with only two Q-band. Excited the synthesized complexes, the product show the fluorescence spectra, emitted range 652-655 nm for TOMPP and emitted range 645-654 nm for metal complexes. The thermal behaviours including the possible phase transition of porphyrins were studied during the heating process at lower temperature (298-973 K). The CuTOMPP exhibit lower thermal stability compared with TOMPP. However, free base ligand and metalloporphyrins are stable enough for real applications, especially in ITO coating process for solar cell.

References

1. Messerschmidt A., Huber R., Poulos K., Wieghardt K., *Handbook of metalloproteins* (2001)
2. Bertini I., Gray H.B., *Biological Inorganic Chemistry*. University Science Books, Sausalito, (2007).
3. R.G. Little, J. Heterocycl. Chem. **18**, 129 (1981)
4. T. Mashiko, J.C. Marchon, D.T. Musser, C.A. Reed, J. Am. Chem. Soc. **101**, 3653 (1979)
5. W.D. Woggon, Acc. Chem. Res. **38**(2), 127-136 (2005)
6. E. Brulé, K.K. Hii and Y.R. de Miguel, Org. Biomol. Chem., **3**, 1971-1976 (2005)
7. G. Fagadar-Cosma, E. Fagadar-Cosma, I. Popa, I. Taranu, Chem. Bull., **52**(66), 109-112 (2007)
8. I.D. Kostas, A.G. Coutsolelos, G. Charalambidis, A. Skondra, Tetrahedron Letters., **48**, 6688-6691, (2007)
9. J. Li, H. Xin and M. Li, Liquid Crystals, **33**, 913-919 (2006)

10. T.E.O. Screen, I.M. Blake, L.H. Rees, W. Clegg, S.J. Borwick and H.L. Anderson, J. Chem. Soc., Perkin Trans. **1**, 320-329 (2002)
11. A.V.C. Simoes, A. Adamowicz, J.M. Dąbrowskib, M.J.F. Calvete, A.R. Abreu, G. Stochel, L.G. Arnaut, M.M. Pereira, Tetrahedron., **68**, 8767-8772 (2012)
12. D.E. Dolmans, D. Fukumura, R.K. Jain, Nat. Rev. Cancer., **3**, 380-387 (2003)
13. B. Krammer, Anticancer Res. **21**, 4271-4277 (2001)
14. S.G. Rockson, D.P. Lorenz, W.F. Cheong, K.W. Woodburn, Circulation, **102**, 591-596 (2016)
15. M.T. Brumbach, A.K. Boal, D.R. Wheeler, Langmuir, **25**, 10685–10690 (2009)
16. H. Shargh, A.H. Nejad, Tetrahedron., **60**, 1863-1868 (2004)
17. B. Bandgar, P. Gujarathi, J. Chem. Sci., **120**, 259-266 (2008)
18. M.D. Togashi, M. Costa, N. A. Sobral, A. R. Gonslaves, Chemical Physics, **300**(1-3), 267-275 (2004)
19. A. Boudif, M. Momenteau, J. Chem. Soc. Chem. Commun, 2069 (1994)
20. B. Ventura, L. Flamigni, New J. Chem., **32**, 166-178 (2007)
21. Simple method of measuring the band gap energy value of TiO_2 in the powder form using a UV/Vis/NIR spectrometer, Available form: http://www.perkinelmer/Content/applicationnotes/app_uvvisnirmeasurebandgapenergyvalue.pdf
22. M. Gamboa, M. Campos, The Journal of Chemical Thermodynamics, **42**(5), 666-674 (2010)

MATEC Web of Conferences **69**, 07001 (2016) DOI: 10.1051/matecconf/20166907001
ICCPE 2016

An Optimization Study on Syngas Production and Economic Evaluation

Faraz Qasim, Jae Sun Shin, Jong Hwa Jeong and Sang Jin Park

Deptartment of Chemical Engineering, Dongguk University, 26, Pil-dong 3-ga, Jung-gu, Seoul 100-715, Korea

Abstract. Syngas production in Gas-to-liquid (GTL) process is focused in past by several researchers to increase the production with minimal capital and operating costs. In this study, syngas production process is simulated and optimized to increase its production and the economic analysis is studied for the proposed optimized process. Aspen HYSYS v8.4 is used for all process simulation work in this article. A new configuration is rigorously simulated while using auto-thermal reforming. Results exhibit a tremendous rise in production of syngas.

1 Introduction

The gas-to-liquid (GTL) technology has gained much attraction due to recent rise of crude oil prices. Many energy related companies have shown their interest in gas-to-liquid technology in recent .time and as a result many large scale industrial facilities are producing synthetic oil by using natural gas. The natural gas reserves are being utilized now to be converted to useful liquid products by using GTL process [1]. Franz Fischer and Hans Tropsch developed the FT (Fischer Tropsch) process which acts a key role to convert syngas into range of hydrocarbons [2], [3]. In industry, syngas can be produced by different methods; partial oxidation, dry reforming and steam methane reforming while the combination of two or more aforementioned processes can also be utilized for production of syngas e.g autothermal reforming (ATR), tri-reforming, gas heated reforming (GHR) [4]. The raction rate and the overall conversion is higher in ATR than the steam reforming process due to that fact that the exothermic combustion reactions occur in the ATR reactor [5-7]. A mixture of CO, H2, CO2 with traces of some unconverted components are obtained as a result of all reforming technologies. ATR technology is most used method as it produces the H2/CO ratio of 2. Hence, the ratio produced and process conditions match well with the FT reactor. The overall reaction rates in ATR are also better than other reformer types. ATR process is usually considered favorable for production of synthesis gas. Synthesis gas in the desired range can be produced by using ATR. Moreover, the floating production facilities prefer ATR due to its compact size. In recent time, ATR technology is being operated by various license holders like Haldor Topsie and Linde.

The other useful process for producing synthetic gas is stream reforming which has a long history of being used in process industries. It has advantage over other processes that it uses steam for production of carbon monoxide, carbon dioxide and hydrogen. The other processes consume oxygen as an external utility for conversion of natural gas and the use of oxygen leads to environmental hazards. The use of technologies like ATR and partial oxidation also leads to carbon deposition, safety issues and catalyst destruction [8]-[10]. Another advantage of steam reforming can be in the form of hydrogen produced in the process which can be further consumed in the upgrading step of GTL process.

2 Process description

Autothermal reforming of natural gas produces range of syngas composition due to which it is considered useful in the gas to liquid applications. An improved design of autothermal reforming process is proposed and its optimization is done to make it more economical. The production of syngas is increased and two cases are compared on the basis of different configurations.

The partial oxidation and steam methane reforming reactions [11, 12] occur in ATR process. The reactions are shown below:

$$CH4 + 1.5\ O2 \leftrightarrow CO + 2\ H2O \quad (1)$$

$$CH4 + H2O \leftrightarrow CO + 2\ H2 \quad (2)$$

$$CO + H2O \leftrightarrow CO2 + H2 \quad (3)$$

$$CH4 + 2\ H2O \leftrightarrow CO2 + 4\ H2 \quad (4)$$

The Kinetics applied for the catalytic reforming reactions are taken from Xu and Froment [13]. The simulation of the syngas production process is run by Aspen HYSYS v8.4. Fig. 1 depicts the process simulation of the syngas production process with recycling technique and heat integration. While Fig. 2 shows the conventional design for synthesis gas production. For simulating the process, Peng-Robinson [14] property package is utilized in Aspen HYSYS

DOI: 10.1051/matecconf/20166907001

simulator. In the proposed design (Fig. 1), natural gas and steam are introduced at temperature 400 °C and 300 °C, respectively, while the pressure in this design is 25 bar. PFR-100 is reformer, while X-100 and X-101 are hydrogen removal unit and CO removal unit, respectively. For reactor, PFR model is utilized and the product stream from reactor leave at temperature 715 °C and pressure 2500 kPa. A process to process heat transfer method is also utilized in this design to make this process optimized unlike previously proposed designs in literature. Hence, temperature for stream 3 and oxygen stream is raised up to 598° C and 200° C, respectively. After the partial oxidation, natural gas is introduced to reformer where it is converted to syngas. In this work, conventional GTL process is coupled with the recycling technique which includes recycling of tail gas to auto-thermal reformer due to which the molar flows of the syngas are increased while the H2/CO ratio still remains same as 2.

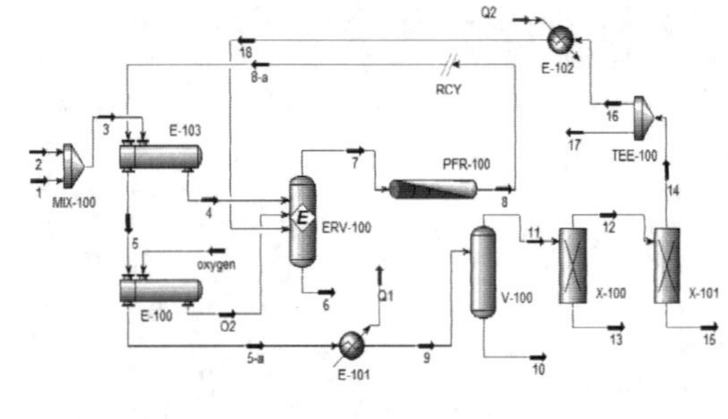

Figure 1. Autothermal reforming process for syngas production with tail gas recycled (optimized design)

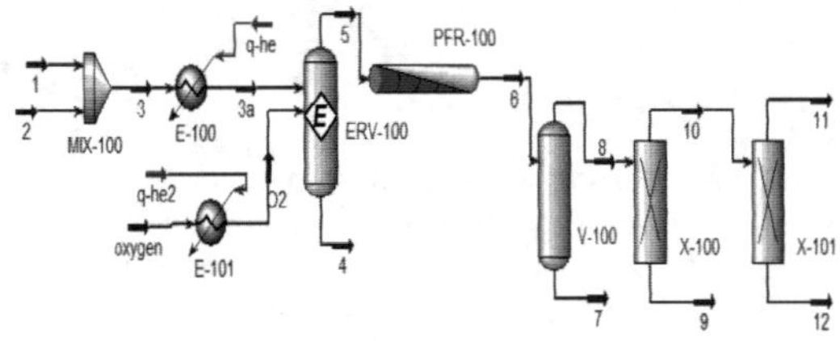

Figure 2.Autothermal reforming process for synthesis gas production with conventional design

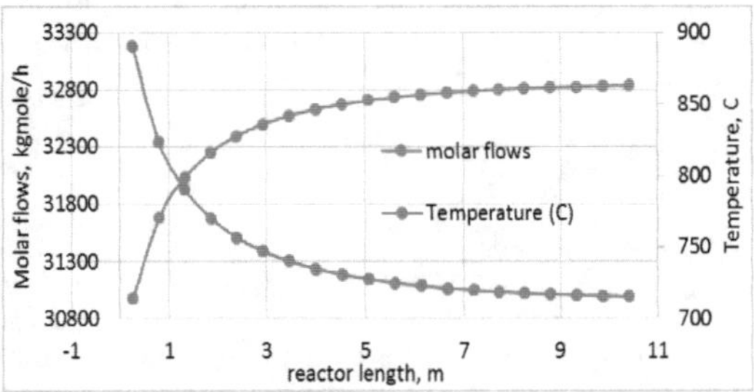

Figure. 3. Reactor length vs molar flows and temperature

In proposed design (Fig. 1), Hydrogen is recovered as bottom product of X-100 unit while the carbon dioxide is recovered in the bottom of X-101 unit. The product stream from the combustion reactor leaves at 1109 °C and it enters PFR. The conversion of methane in PFR results into 98% and the product stream from PFR is at high temperature i.e 715.6 °C. A different configuration is proposed to cool down this stream and two heat exchangers are utilized in the process. The process with and without heat exchangers are simulated and the economic aspects of both processes are analyzed. The recycled tail gas mainly constitutes unconverted methane, hence, the recycling provides an increased production in an optimized design. The recycling technique is proposed to increase the volumetric production of syngas. A conventional process is also designed and its production and operating costs are analyzed.

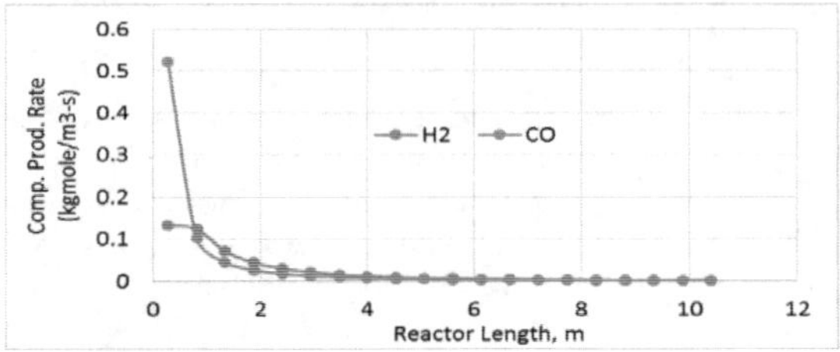

Figure. 4. Reactor length vs comp. prod. rate

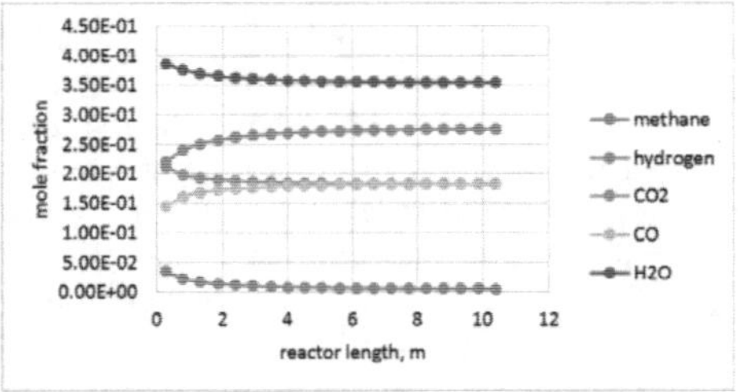

Figure 5.Reactor length vs compositions of components.

It can be observed that the conventional design is utilizing external hot steam to meet the energy requirements. Hence, the capital cost and operating cost of the system get increased while using conventional design for syngas production by autothermal reforming process. Feed temperature is increased because of the high temperature requirement in the reactor. Oxygen feed is also heated upto 200 °C for process requirement.

3 Results and discussions

The desired composition of syngas is produced and optimization is done for whole process. Fig. 3, 4 and 5 show the relation of molar flows, temperature, reaction rates and mole fractions of different components with varying reactor length. Fig. 3 explains the relation of the temperature and molar flows along the reactor length. The temperature along the reactor decreases as the length of the reactor increases. While the molar flows show a direct relation towards the reactor length. Hence, the reactor length plays an important role for the operating conditions of the process. Fig. 4 exhibits the relation of the product component rates along the reactor length.

The optimization work is done to optimize the operating conditions of the process. Furthermore, the fixed capital cost can also be decreased by using optimization techniques. The small reactor length shows the high flow rates for the product components while increasing the reactor length will results into less flow rate of the product components. As the process deals with components like hydrogen, carbon dioxide, methane, corbon monoxide and water, there is a need to optimize the reactor length for all these compoents. The Fig. 5 demystifies the effect of reactor length on compositions of all the componets dealt in the process. Hydrogen and carbon monoxide exhibit a direct relation to the reactor length while the other components' composition get decreased as the reactor length increases. Table 1 shows optimziation work for the effect of reactor length on temperature and Table 2 exhibits the simulation results for the compositions of all components in reactor.

Table 1. Optimization work for reactor length vs temperature

Reactor length (m)	methane	hydrogen	CO_2	CO	H_2O
0.26	3.50E-02	0.22	0.21	0.14	0.38
0.80	2.32E-02	0.24	0.19	0.16	0.37
1.33	1.74E-02	0.25	0.19	0.16	0.36
1.86	1.40E-02	0.25	0.19	0.17	0.36
2.40	1.17E-02	0.26	0.18	0.17	0.36
2.93	1.01E-02	0.26	0.18	0.17	0.36
3.46	8.93E-03	0.26	0.18	0.17	0.35
4.00	8.04E-03	0.26	0.18	0.17	0.35
4.53	7.35E-03	0.27	0.18	0.18	0.35
5.06	6.80E-03	0.27	0.18	0.18	0.35
5.60	6.37E-03	0.27	0.18	0.18	0.35
6.13	6.03E-03	0.27	0.18	0.18	0.35
6.66	5.75E-03	0.27	0.18	0.18	0.35
7.20	5.52E-03	0.27	0.18	0.18	0.35
7.73	5.33E-03	0.27	0.18	0.18	0.35
8.26	5.18E-03	0.27	0.18	0.18	0.35
8.80	5.05E-03	0.27	0.18	0.18	0.35
9.33	4.95E-03	0.27	0.18	0.18	0.35
9.86	4.86E-03	0.27	0.18	0.18	0.35
10.40	4.79E-03	0.27	0.18	0.18	0.35

Table 2. Compositiions for all components vs reactor length

Reactor Length (m)	0.26	0.80	1.33	1.86	2.40	2.93	3.46	4.00	4.53	5.06	5.60
Temperature (C)	890.23	823.26	790.23	770.17	756.79	747.31	740.30	734.97	730.86	727.61	725.03

The heat supplied from external sources is optimized to make the process energy efficient. The process to process heat transfer shows promising results in optimization procedure. The heat exchangers E-103 and E-100 are utilized in simulation of whole process to exchange heat with the process streams. Hence, the external steam requirements are tremendously decreased. The process to process heat transfer and heat integrations have been studied by various researchers in past literature [15], [16]. The heat integration can be done by using heat pumps or by using process to process heat transfer. A heat pump can be added to the system which increases the heat quality provided to reboiler. The heat pump can be absorption heat pump or mechanical heat pump. Mechanical heat pump makes use of vapor recompression technique which is very efficient way to save external steam and hot utilities. The product stream pressure can be increased by the compressor which provides heat to the process streams. But the use of compression system increases the overall cost of the system and hence it makes process economically unattractive. Hence, compression system is not used in this work and the heat exchangers are utilized to exchange the heat within the process.

The heat exchanger E-103 is utilized to increase the temperature of the feed stream from 348.1°C to 598.1°C and this heat stream 4 is then fed to the reactor. While the other heat exchanger E-100 is utilized to heat up the oxygen feed from 20°C to 200°C. The stream 5 which still has the high heating potential is then used to heat up oxygen feed in E-100. As a result, temperature of stream 5-a is decreased.

In this work, the production rate of syngas in the designed process is found out to be approximately 11570 kgmol/h. After sizing and mapping of all equipments, economic evaluation is also done and objective function is defined to analyze economics.

An objective function is defined to analyse the economic aspects of both designs and the profit. The mathematical formulation representing objective function is given in equation 5.

$$\text{Objective function} = PR - RM - Ut - CD \quad (5)$$

Product revenue (PR), described in equation 5, is the price of the product recovered in the product stream and it is defined in equation 6.

$$\text{Product revenue} = \sum P_{product} F_{product} \quad (6)$$

where P is the price of the product in product stream and F is the flow rate of the product. Raw material RM cost is used in the objective function and it is defined in equation 7.

$$\text{Raw material cost} = \sum C_{feed} F_{feed} \quad (7)$$

Equation 7 defines raw material cost as the product of cost of feed components and the flow rate of feed components.

Furthermore, Utility cost is another important factor which is considered in the economic analysis. The utility cost is cost of the steam and cold water used in the process. As the optimization work is done in this simulation study, cost of steam is optimized and the requirement of steam and cold water is decreased as much as possible. The utility cost is defined in equation 8.

$$\text{Utility Cost} = \sum C_{Utility} \quad (8)$$

Equation 5 makes use of capital depreciation CD which is defined as the product of the capital cost of the equipment and the depreciation factor. The depreciation factor is found from the Aspen HYSYS economic analyzer and the CD is found by using following equation.

$$\text{Capital depreciation} = \sum C_{equipment} D \quad (9)$$

Where C is the cost of the equipment and D is depreciation factor.

Economic evaluation of the syngas production process is done by using Aspen HYSYS Process Economic Analyzer. First, all the equipment are sized and mapped according to the process conditions. Economic analysis is done for the optimized processes and the optimized variables are first fixed in the simulation flow sheets.

The conventional process exhibits the project capital cost of 1.96E+06 USD and operating cost of 1.44E+06 USD/year. The total utilities consumed in this design cost up to 4.9E+05 USD/year. Electricity requires 4.05 USD/hour while steam needed in the process costs 51.72 USD/hour. The optimized process (Fig. 1) is the solution for high steam cost and this design does not use external hot steam for process requirements. The economic analysis shows that the conventional process is very expensive than the optimized model using recycling and heat integration. The project capital cost and operating cost of optimized design (Fig. 1) is found to be 1.26E+06 USD and 9.14E+05 USD/year, respectively. The desired rate of return in this design is found out to be 20 percent/year. For calculating capital depreciation, the depreciation factor found in economic analysis is 2.5E+05 per year. The depreciation factor is calculated by straight line method.

The maintenance cost and operating labor cost are the major factors which are needed in economic analysis. In the optimized process, the operating cost and the maintenance cost are found to be 1.94E+03 USD and 4.82E+05 USD, respectively. Moreover, the insulation cost, contingency and the instrumentation costs are calculated to be 2.30E+04, 2.45E+05 and 1.61E+05 USD, respectively.

As the heat integration has also been applied in the process, there is no usage of external steam to heat up the streams. All the heating requirements are met within the process by process to process heat transfer. Hence, the external utility includes only electricity and cooling water. Electricity consumption costs 4.05 USD/hour while the cooling water costs 785.82 USD/hour in the designed process for syngas production. The objective function is used to compare the profit of the process with and without heat integration and the process with heat integration exhibits more profit.

In this study, aspen HYSYS simulations of different configurations, optimization work and economic analysis are performed. The optimized model using recycling techniques and heat integration exhibit 14% more profit than the process without recycle streams and heat integration. Hence the operating cost can be decreased by the proposed design in this simulation study.

4 Conclusions

A design for autothermal reformer for GTL process is developed alongwith optimizations to make the process economically favorable. Simulation of conventional and optimized design is done for syngas prodcution. Tail gas recycling configuration gives the most optimized process design with H2/CO ratio of 2. The recycling method used in this study increases the production of the syngas. Process to process heat transfer is also done to optimize the utility requirements and the results show that the proposed model requires less utilities than the designs proposed in previous literature. The economic aspects of the process are utilized too and it unveils the total operating cost and total capital cost of the process. The proposed design using recycle technique and heat integration exhibits 14% more profit than the conventional design. Moreover, the volumetric production of syngas has been improved in this work.

References

1. H. Schulz; Short history and present trends of Fischer-Trosch synthesis, Applied Catalysis A, **186**, 3–12 (1999).
2. M. E. Dry; The Fischer Tropsch Process: 1950-2000, Catalysis Today, **71**, 227–241 (2002).
3. T. Takeshita, K. Yamaji; Important roles of Fischer-Tropsch synfuels in the global energy future, Energy Policy, **36**, 2773– 2784 (2008).
4. A. P. E. York, T. C. Xiao, M. L. H. Green, J. B. Claridge,; Methane oxyreforming for synthesis gas production, Catal. Rev. Sci. Eng., **49**, 511–560 (2007).

5. X. Song, Z. Guo; Technologies for direct production of flexible hydrogen/CO synthesis gas, Energy Conversion and Management, **47**, 560–569 (2006).
6. K. Aasberg-Petersen, T. S. Christensen, C. S. Nielsen, I. Dybkjaer,; Recent developments in autothermal reforming and pre-reforming for synthesis gas production in GTL applications, Fuel Process. Technol., **83**, 253–261 (2003).
7. Y. Patcharavorachot, M. Wasuleewan, S. Assabumrungrat,
A. Arpornwichanop; Analysis of hydrogen production from methane autothermal reformer with a dual catalyst-bed configuration, Theor. Found. Chem. Eng., **46**,658–665 (2012).
8. Padban, N.; Becher, V. Clean Hydrogen-Rich Synthesis Gas ;Chris Gas Report, (2005).
9. Zhu, Q.; Zhao, X.; Deng, Y. Advances in the Partial Oxidation of Methane to Synthesis Gas.J. Nat. Gas Chem., **13**, 191 – 203 (2004).
10. Van Hardeveld, R. M.; Groeneveld, M. J.; Lehman, J. Y.; Bull, D. C. Investigation of an Air Separation Unit Explosion. J. Loss Prev. Process Ind., **14**, 167 – 180 (2001).
11. W. Maqbool & E. S. Lee, Syngas Production Process Development and Economic Evaluation for Gas - to - Liquid Applications. Chemical Engineering & Technology, 37, 995-1001 (2014).
12. J. Pina, D. O. Borio; Modeling and simulation of an autothermal reformer, Latin Am. Appl. Res., **36**, 289–294 (2006).
13. J. Xu, G. F. Froment; Methane steam reforming, methanation and water-as shift: I. Intrinsic kinetics, AIChE J., **35**, 88–96 (1989).
14. J. N. Jaubert; F. Mutelet. VLE predictions with the Peng-Robinson equation of state and temperature dependent calculated through a group contribution method. Fluid Phase Equilibria, **224**, 285-304 (2004).
15. F. Qasim, J. S. Shin, S. J. Cho, & S. J. Park, Optimizations and heat integrations on the separation of toluene and 1-butanol azeotropic mixture by pressure swing distillation. Separation Science and Technology, 51, 316-326 (2016)
16. T. O. Omideyi; J. Kasprzycki, F. A. Watson, The economics of heat pump assisted distillation systems I. A design and economic model. J Heat Recov Syst., **4**, 187 –200 (1984).

MATEC Web of Conferences **69**, 08001 (2016) DOI: 10.1051/matecconf/20166908001
ICCPE 2016

Green, Eco, Innovative Design, and Manufacturing Technology of a 1-Ton per Batch Municipal Solid Waste Incinerator

Somrat Kerdsuwan

The Waste Incineration Research Center, Department of Mechanical and Aerospace Engineering, Faculty of Engineering, Science and Technology Research Institute, King Mongkut's University of Technology North Bangkok, Bangkok, Thailand

Abstract. The thermal treatment of waste by incineration is considered an ultimate solution in order to get rid of waste properly by using the combustible properties of waste and transforming them into inert form and gaseous emission, with the main advantage of a huge reduction in mass and volume of treated waste, destruction of the dangerous components in waste, and obtaining green and clean energy from the exothermal reaction from the completed combustion process. In order to achieve the main goal of incineration, a good design, construction, supervision, and intensive operation and maintenance must be taken into account, especially for the small-scale incinerator. This research will deal with the green, innovative, and eco design and manufacturing technology of a 1-ton per batch municipal solid waste (MSW) incinerator. The concept design of the incinerator will focus on the design of the feeding process where only one batch of waste will be discharged into the combustion chamber at one time instead of the semi-feed process, as found in the conventional incinerator. This will ease the operation of the operator and reduce the operating cost. Moreover, the innovative design includes the redesign of combustion air injection into either the primary or secondary combustion chamber in order to achieve the 3Ts of combustion (time, temperature. and turbulence). This design can eliminate the use of an auxiliary burner in the primary combustion chamber. Rethinking the innovative design of using recirculation hot flue gas for preheating of wet garbage in order to pre-dry the waste before combustion is also taken into account. The manufacturing process of the wall composition as well as other parts of the incinerator are also examined.

1 Introduction

The disposal of municipal solid waste (MSW), which is the discard of matter from human activities, must be done in a proper manner in order to avoid the impact on the environment. Regarding the small community in urban areas, landfill cannot be implemented for the disposal of waste due to the scarcity of land. Onsite treatment by incineration is one of the promising options to treat this waste [1]. However, due to the small capacity of facilities, the burning of waste does not yield enough by-product energy to produce heat or electricity, which causes high operating cost since additional fuel is required to achieve a high destruction temperature [2]. Moreover, with the strengthening emission standards nowadays, many onsite incinerators remain unused [3]. The objective of this work is to redesign and manufacturing of the Green, Eco, Innovative Design and Manufacturing technology of a 1-ton per batch MSW in order to use the incinerator at the source of MSW generation. The operating batch time was about 6 to 8 hrs and needed only a one-time charge of MSW into the combustion chamber by using a dumping garbage truck. Only one operator was needed to take care of the system. It was designed to achieve inexpensive operating costs but would be able to attain emission standards which would lead to a Green, Eco and Innovative small-scale MSW incinerator.

2 General Concept of the Innovative Design of a 1-ton per batch MSW Incinerator

The general operating principle of the small-scale MSW incinerator consists of charging of MSW in the combustion chamber in a batch process. Each batch is enough to partially burn for 10 to 15 minutes, then the next charging will start to feed the MSW in the chamber again. Advantage of this technique is to help the combustion to be taken place in a short time and can avoid the excess pollutants coming from the primary combustion chamber (PCC) to the secondary combustion chamber (SCC). However, one drawback is that it is a time-consuming process and requires an operator to take care of the incinerator at all times. As a result, an innovative design is to try to avoid this too-frequent feeding time on the incinerator.

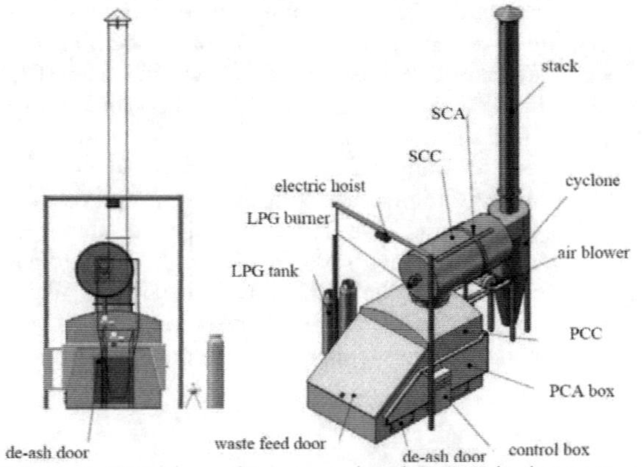

Figure 1. Overview of 1-ton per batch MSW incinerator.

The next consideration is the high operating cost due to the use of an additional fuel burners in the primary combustion chamber. The purpose of this burner is to sustain the desired combustion temperature in the chamber when burning high-moisture waste content, especially with wet fraction MSW (food waste, vegetation waste, etc.) The more that this additional burner is used, the more additional fuel must be used (diesel oil, for example), which results in high operating costs. In order to overcome this, the less use or even no use of the burner muse be taken into account, even by improving the combustion process by using a novel design of combustion air distribution or by mixing the characteristics of waste and air.

In order to achieve complete combustion, not only must the proper ratio of waste and air be matched, but the 3Ts concept of combustion, including Time, Temperature and Turbulence must be implemented during the design of the incinerator [4]. Time ensures enough reaction time between the waste and air, both in the PCC or SCC. The time in the PCC will deal with the solid-phase combustion of the waste and primary combustion air (PCA); it must be long enough to burn all of the carbon contained in the waste, so only a small fraction of carbon will be left in the ash. The **Time** in the SCC concerns the gas-phase combustion of the flue gas coming from the PCC, which may contain an incomplete combustion product, and this must be re-burnt with secondary combustion air (SCA) and have enough time for complete combustion. **Temperature** is a control combustion parameter both for the PCC and SCC. The temperature in the PCC must be high enough for completely burning the solid fraction of waste and for transforming it completely into ash. In the SCC, the temperature must be high enough to ensure complete combustion of the flue gas. It is recommended that, for the MSW incinerator, the temperature in both the PCC and SCC must be at least 850°C [5]. **Turbulence** is a mixing characteristic in the combustion chamber in order to ensure complete combustion. In the PCC, turbulence can be performed in terms of the movement of the waste in the chamber or the injection location of the PCA in the PCC, where in the SCC, turbulence means the mixing of the flue gas coming from the PCC and SCA. The degree of mixing in the PCC and SCC can be investigated through the study of the computational fluid dynamic (CFD) [6].

An overview design of the 1-ton per batch MSW incinerator in this study is shown in Figure 1. It consists of three main parts: the PCC, SCC, and cyclone. The MSW will be fed into the PCC through the feed gate which can be opened by an electric hoist. It is charged one time into the PCC and the feed gate will close. The PCA is fed into the PCC through the air box around the body of the PCC. Hot flue gas from the PCC will flow up to the SCC, which is located on the top part and re-burned with the SCA, which is fed at the circumference of the SCC in order to increase the degree of mixing. The products of the complete combustion will flow out from the SCC and lead to a cyclone for dust and particulate collection. The cyclone which was selected for use in this study is only one option for controlling dust and particulate. Other alternatives for air pollution control strategy such as a wet scrubber, a bag filter, or an electrostatic precipitator can be implemented based on emission regulations. The designed system uses only one additional burner located at the SCC to ensure the desired combustion temperature in the SCC. No burner is used in the PCC in order to reduce the operating cost of the incinerator. Additional fuel used in this design is LPG, which costs half the price of diesel oil.

3 Designed Waste Characteristic

The characteristic of the MSW used for this study came from the average values of developing country's MSW characteristics [1] which shown in Table 1 :

Table 1. Design waste characteristic

Moisture: 32%	Ash: 7%
Heating Value: 14,700 kJ/kg (dry basis)	Bulk Density: 200 kg/m^3

4 Innovative Design and Manufacturing of feeding system

In order to design the charging of the MSW into PCC in one batch, it is recommended to use the charging of waste from a dump garbage truck returning from collecting waste from households. The MSW will be dumped into the PCC at one time. Since the desired capacity of the dumping is one ton per batch with a bulk density of waste of 200 kg/m^3, the MSW volume in one batch is 5 m^3. However, the volume of the PCC must be at least 2 times the MSW volume in the chamber in order to have enough space for the combustion process, and consequently, the design volume of the PCC will be 10 m^3. Knowing the width of the garbage dump truck, the dimension of the PCC can be determined. The opening of the feed gate is designed as a hinge, which is opened by a one-ton electric hoist. After the charging process completed, it will be closed before the ignition process begins.

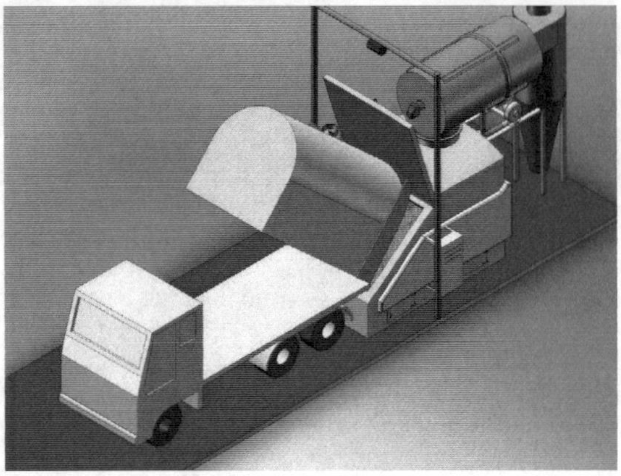

Figure 2. Feeding of MSW into the PCC using a garbage dump truck

5 Innovative Design of PCC and SCC

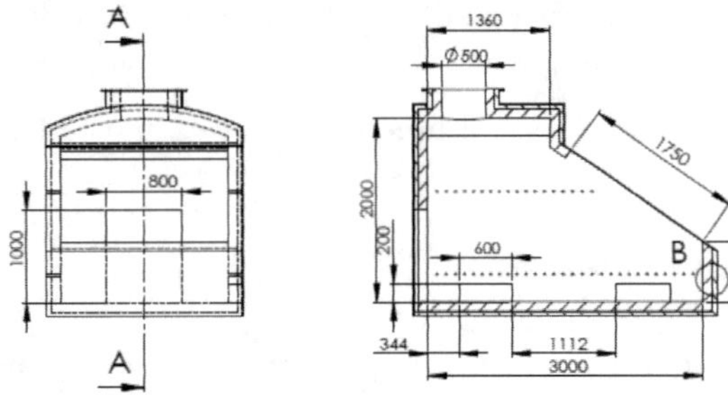

(a)

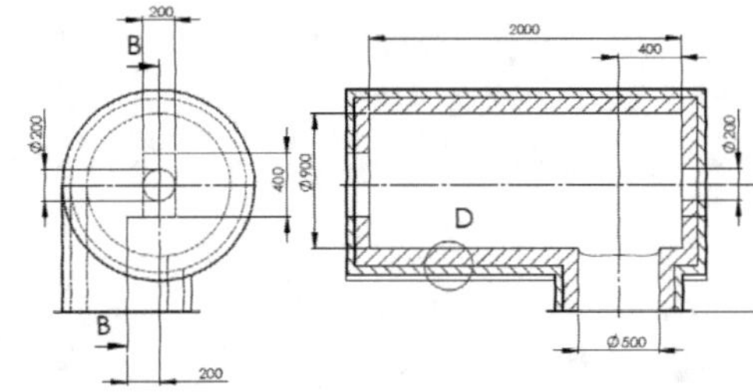

(b)

Figure 3. A design of innovative 1 Ton per day incineration system (units in mm.)

Figure 3 shows the design of the PCC and SCC. Waste will be charged in the PCC and will sit on the grating, which is made of cast iron, in order to sustain a high temperature during the combustion. The grating is placed about 10 cm above the PCC's floor. Ash, which is a result of the combustion, will fall between the spaces of the grating to the floor. The doors on the side of the PCC are used to remove the ash. The PCA coming from an air blower will be distributed and injected into PCC through multi-holes on both sides. There are two levels of air holes; the lower holes are used for combustion of the air of solid waste putting on the grate, and the upper holes are used to provide additional air to burn the volatile matter evaporated from the solid bed. Each row of air holes will have valves to freely control the combustion process and can be regulated by the operator. The SCC is located on the top of the PCC and put in a location so that the flue gas coming from the PCC will flow circumferentially into the SCC in order to increase the degree of mixing to meet the turbulence requirements for complete combustion. The SCA also comes from another air blower and is injected into the SCC as a ring around the circumference of the SCC wall. The volume of the SCC is designed to have a residence time for the gas retained in the chamber longer than 1 second.

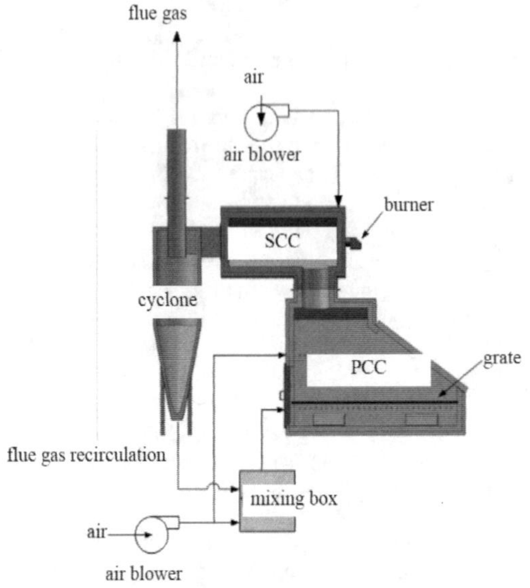

Figure 4. Innovative design for the recirculation of hot flue gas for reburning in the PCC

The designed system uses only one additional LPG burner installed at the SCC. Its function is only to assure the reaching of the desired SCC temperature (850°C) by igniting the LPG as additional fuel. Once it reaches the desired temperature, the burner will stop the injection of the LPG but will feed only combustion air. Energy savings could be accomplished by recirculating the hot flue gas from the cyclone and reverting it to mix with the PCA in the mixing box before being injected into the PCC, and therefore, installation of a PCC burner is not required (Figure 4). Ignition of the MSW in the PCC can be done by igniting it with a match and a small amount of gasoline poured on top of the waste in the PCC. The PCC and SCC wall consists of three layers. The inner layer is made of castable refractory, which can sustain a combustion temperature up to 1,400°C with a thickness of 10 cm. The second layer is a thermal insulator used to keep the heat inside the chamber and can sustain a temperature up to 1,100°C with a thickness of 5 cm. The outer layer is a steel structure with a thickness of 4.5 mm.

The design criteria for the Green, Eco, Innovative Design and Manufacturing Technology of one-ton per batch MSW incinerator can be summarized as follows:

- The destruction capacity of the combustor is one ton per batch for 14,700 kJ/kg (dry basis) heating value of waste
- The dumping of the waste into the incinerator from the garbage truck has been applied for only a one-time feed
- The combustion air in the PCC is divided into upper fire and under fire for complete combustion of the MSW
- Only one additional burner is used at the SCC and for the recirculation of the hot flue gas for reburning
- It is designed for one-time feed of MSW with a batch operating time of 6 to 8 hrs with one operator

6 Summary

A small-scale one-ton per batch MSW incinerator with a green, eco, and innovative design concept and manufacturing technology has been established. The idea is to use this design unit at the source of waste with lower operating and maintenance costs, less need of an operator to operate and control it, but still achieve a good combustion process and less impact on the environment. The one-time feed of waste into the combustion chamber leads the non-use of an auxiliary burner in the primary combustion chamber. The 3Ts concept of complete combustion has been achieved via innovative design and manufacturing.

References

1. K. Laohalidanond, P. Chaiyawong, S. Kerdsuwan, *Municipal Solid Waste Characteristics and Its Green and Clean Energy Recovery in Megacity of Asian Countries*, 2015 International Conference on Alternative Energy in Developing Countries and Emerging Economies, Bangkok, 2015
2. S. Kerdsuwan, *Innovative Design and Manufacturing Technology of High Temperature Air Combustion (HTAC) Municipal Solid Waste Incinerator*, Applied Mechanics and Materials Vols. 789-790 (2015) pp 375-379
3. S. Kerdsuwan, K. Laohalidanond, W. Jangsawang, *Sustainable Development and Eco-friendly Waste Disposal Technology for the Local Community*, 2015 International Conference on Alternative Energy in Developing Countries and Emerging Economies, Bangkok, 2015
4. S. Kerdsuwan, K. Laohalidanond, P. Chaiyawong, *A Novel Hybrid Design of Incineration – Gasification for Energy Saving*, Applied Mechanics and Materials Vols. 799-800 (2015) pp 95-99
5. C. R. Brunner, *Incineration Systems Handbook*, McGraw-Hill, (New York, NY), 1991
6. P. Meenaroch, S. Kerdsuwan, K. Laohalidanond, *Development of Kinetics Models in Each Zone of a 10 kg/hr Downdraft Gasifier by Using Computational Fluid Dynamics*, 2015 International Conference on Alternative Energy in Developing Countries and Emerging Economies, Bangkok, 2015

AUTHOR INDEX